Endophytes of the Tropics

Endophytes of the Tropics

Diversity, Ubiquity, and Applications

Adeline Su Yien Ting

CRC Press
Taylor & Francis Group
Boca Raton London New York

CRC Press is an imprint of the
Taylor & Francis Group, an **informa** business

First edition published 2021
by CRC Press
6000 Broken Sound Parkway NW, Suite 300, Boca Raton, FL 33487-2742

and by CRC Press
2 Park Square, Milton Park, Abingdon, Oxon, OX14 4RN

© 2021 Taylor & Francis Group, LLC

CRC Press is an imprint of Taylor & Francis Group, LLC

Library of Congress Cataloging-in-Publication Data

Names: Ting, Adeline Su Yien, author.
Title: Endophytes of the tropics : diversity, ubiquity and applications / Adeline Su Yien Ting.
Description: Boca Raton, FL, USA : CRC Press, 2020. | Includes bibliographical references and index.
Identifiers: LCCN 2020014067 (print) | LCCN 2020014068 (ebook) | ISBN 9780367508876 (hardback) | ISBN 9780429061387 (ebook)
Subjects: LCSH: Endophytes. | Tropical plants--Microbiology.
Classification: LCC QR351 .Y546 2020 (print) | LCC QR351 (ebook) | DDC 581.70913--dc23
LC record available at https://lccn.loc.gov/2020014067
LC ebook record available at https://lccn.loc.gov/2020014068

ISBN: 9780367508876 (hbk)
ISBN: 9780367184063 (pbk)
ISBN: 9780429061387 (ebk)

Typeset in Times
by Lumina Datamatics Limited

To my family, for their love, affection, and understanding, and above all, for showing me that all things are possible

Contents

Foreword ix
Preface xi
Acknowledgment xiii
Author xv
Introduction xvii

1 **Endophytes in the Tropics** **1**
1.1 Diversity and Ubiquity 1
1.2 Endophytes and Their Tropical Host Plants 4
1.3 Factors Driving Endophyte Communities in the Tropics 14
1.4 Spread of Tropical Endophytes and Their Endophyte-Host
Plant Association 15
1.5 Endophyte Research in the Tropics 18

2 **Applications of Endophytes from the Tropics** **21**
2.1 Introduction 21
2.2 Tropical Endophytes for Agriculture 22
2.3 Tropical Endophytes for Biomedicine and Biosynthesis
of Valuable Compounds 28
2.3.1 Antimicrobial Compounds 28
2.3.2 Anticancer Compounds 30
2.3.3 Functional Polysaccharides 36
2.3.4 Other Valuable Compounds/Metabolites 36
2.3.5 Biotransformation of Valuable Compounds 39
2.3.6 Biosynthesis of Nanoparticles by Tropical
Endophytes 42
2.4 Tropical Endophytes for Industrial Use 43
2.4.1 Enzymes from Tropical Endophytes 43
2.4.2 Pigments (or Colorants) from Tropical Endophytes 45
2.5 Tropical Endophytes for Bioenergy and Biocatalysis 48
2.6 Tropical Endophytes for Bioremediation 49
2.6.1 Metal Bioremediation 49
2.6.2 Hydrocarbon Bioremediation 54
2.6.3 Other Xenobiotic Pollutants 55

3 Valuable Endophytic Species from the Tropics   **57**
   3.1   Endophytic Actinobacteria   57
   3.2   *Diaporthe and Phomopsis* Species   58
   3.3   *Lasiodiplodia* Species   59
   3.4   *Muscodor* Species   60
   3.5   *Trichoderma* Species   60
   3.6   *Xylaria* Species   61

4 Commercialization of Endophytes from the Tropics   **63**
   4.1   Introduction   63

5 Challenges in Endophytic Research in the Tropics   **65**
   5.1   Challenges and Limitations   65
   5.2   Remedial Strategies to Challenges   67

6 Conclusions   **71**

References   73
Index   105

Foreword

The presence and existence of endophytes in the tropics is a blessing. They have been associated with the many applications that are beneficial to mankind and the environment. The research and information on tropical endophytes is, however, rather dispersed, focusing either on diversity, or metabolites, or just the ecological interactions of the endophytes with their host plants.

This book, therefore, aims to provide a more consolidated outlook on the important aspects of endophyte research in the tropics. There will be discussions on their diversity and ubiquity, in relation to the tropical host plants. There will also be discussions on the applications of endophytes, particularly based on the beneficial compounds/metabolites they produce. Highlights on the challenges of endophyte research and some of the strategies to manage the challenges are presented as well. In short, tropical endophytes are interesting and warrant documentation. They have the ability to tolerate and adapt to stressful tropical conditions (and diseases), thus are of great interest. The tropical regions are also rich with plant diversity and ethnobotanical history, thus it makes great sense to further explore and document endophytes from tropical plants. As such, their presence in the tropics is truly appreciated.

Preface

For many of us, the term "endophyte" is foreign. A quick look up on the term will reveal that "endo" means "internal or within," and "phyta" means "plants." As such, endophytes refer to a group of microorganisms that reside or spend part of their lifecycle inside plants. With the term coined, it is clear that plants are vessels that host a diverse multitude of endophytes that may in one way or another impact the plant, environment, and mankind.

In the early days, endophytes were primarily recognized for their role as herbivore deterrents. This role shifted when more and more of these endophytes were isolated and cultured independently from host plants, and were revealed to have other properties. One of the greatest discoveries of endophytes is the production of taxol, first from endophytes of the *Taxus* yew tree, and gradually from other endophytes from other host plants. The discovery of taxol, an anticancer agent, presented a feasible approach to harness taxol for medical purpose. This also meant lesser trees will be claimed for production of taxol. There lies the attractive use of endophytes as sustainable and renewable sources of valuable compounds. In the subsequent years, endophytes have been further explored and manipulated for their roles as plant growth promoters, biocontrol agents, and producers of various enzymes, pigments, biofuel, and in biotransforming new compounds.

SO, WHY TROPICAL ENDOPHYTES?

Endophytes, as literature would reveal, are not exclusive to the tropics. But the endophytes from the tropics are of greater interest following several reasons. Firstly, the tropics host a greater number of flora diversity, thus with more diverse plants, it is expected that there will also be greater corresponding diversity of endophytes. We have reports on endophytes from rainforest trees to medicinal shrubs, including orchids and Rafflesiaceae. Secondly, the conditions in the tropics present predisposing factors that stimulate adaptations in endophytes. The tropics have both biotic (pathogens, insects, pests) and abiotic stress (pH, salinity, temperature, UV radiation, and perturbations), which triggers responses from the endophytes. And very often, part of the response mechanisms would include the production of metabolites. Therefore, endophytes from the tropics are capable to produce an array of metabolites that may have different functionalities. There is, therefore, a greater possibility that novel metabolites can be isolated from endophytes from the tropics. Thirdly, tropical regions are also rich with ethnobotanical knowledge. It is an unspoken understanding that almost every ethnic tribe in the tropical regions have a list of plants that are applicable to address almost any form of illness or disease. As such, there is a higher probability of finding and harnessing endophytes with desirable biomedicine properties when careful selection of the host plant is made. This is further driven by the hypothesis that endophytes have the capacity to produce similar metabolites as the host plants.

The objective of this book is to showcase the consolidated information on endophytes from the tropics. While the literature may not be exhaustive, the aim is to provide a glimpse of what are some of the endophytes in the tropics, their host plants, their applications in the various field, and some of the challenges in endophytic research. It presents an overview of where we are in endophyte research in the tropics, and how we can progress further into the future. This field of research is evolving and with the advent of technologies, more information would be revealed in the years to come, to bridge the knowledge gap of this amazing world of endophytes.

Acknowledgment

I would like to sincerely thank my research team for their active participation and collaborative effort in making discoveries on endophytes together. We have indeed come a long way. Since the days of morphotyping of endophytes to metagenomics, from the days of plating and enumerating colonies, to qPCR and real-time PCR assays. We have moved with times yet remained steadfast in our pursuit of this niche research area. In remembrance, my thanks too, to my professors and the PROMUSA banana community, who first triggered my interest to explore this wonderful world of endophytes 20 years ago.

My thanks to my friends, whose enthusiasm and encouragement in getting the book published surpasses mine.

And finally, to my family, who knew that science has always been my passion, and writing has always been my pleasure.

This is dedicated to all of you.

Author

Adeline Su Yien Ting holds a PhD in plant pathology from Universiti Putra Malaysia. Her work highlighted the significant value of endophytes as biocontrol agents to suppress Fusarium wilt development in banana plants. Upon graduation, she continued her research on endophytes, diversifying into expounding the potential of endophytes in many areas; for metabolite production, biofungicide application, bioremediation treatment, and biosynthesis of metal-nanoparticles. She leads her applied microbiology research team, and they pursue this interest together. In recent years, she has blended the omics technologies, physical approaches, and instrumentation techniques, such as metagenomics, next-generation sequencing, electron microscopy, Fourier-transformed infrared microscopy, into her research.

Adeline is now with Monash University Malaysia, and is the head of the discipline (Biological Science). She has been awarded 12 external grants, published over 50 journal articles, 10 invited book chapters, and more than 60 conference papers. She has hosted international interns from the RISE-DAAD program and from SupBiotech Paris. The numerous awards conferred have recognized her excellent research work. This includes the prestigious National Outstanding Researcher Award (Biological Science) (2018), Gold Medal for the International Invention, Innovation and Technology Exhibition (ITEX) (2018), ProVice Chancellor Award (Research Excellence) for Postgraduate Research Supervision (2019), and ProVice Chancellor Award (Research Excellence) for Early Career Researcher (2013).

Introduction

Endophytes are microorganisms, which are found inside plant tissues. They comprise bacteria, fungi, and actinobacteria. Endophytes are an interesting group of microorganisms, primarily due to their many definitions over the years. De Bary first described them in 1866 as "any organisms occurring within plant tissues" (Griffin and Carson, 2015). In 1986, Carrol discovered endophytes as organisms that could cause asymptomatic infections in host plants. In the later years, Petrini (1991) refined the term "endophytes" to refer to "all organisms found inhabiting plant organs at some point of their life cycle, without causing any harm to host." As more discoveries were made, more is understood as well of the nature and behavior of endophytes. This includes their ability to remain dormant or latent within plant tissues prior to causing disease, which then revealed that pathogens are potentially mutualistic endophytes (Bashan and Okon, 1981; Schulz and Boyle, 2006). As such, endophytes are deemed to be capable to form symbiotic, mutualistic, antagonistic or pathogenic associations with their host plants, although the presence of endophytes in host tissues is typically symptomless (Petrini, 1991). It has also been postulated that endophytes may alternate between an endophytic and saprophytic lifestyle. This theory arises from observations that endophytes could continue to exist in fallen leaves or leaf litters as saprotrophs (Unterseher et al., 2013). This interchange between endophytic and saprophytic phases has also been observed in root endophytes, which further confirms the many possible lifestyles of endophytes expressed at various stages (Zuccaro et al., 2011).

It is theorized that all plants have plant-endophyte associations. This has been discovered in large trees to herbaceous plants to grasses, as endophytes can colonize and inhabit the various plant tissues, i.e. leaves, branches, stem, and even flowers (Azevedo et al., 2000; Huang et al., 2001). Evidence of endophyte-host coexistence, coevolution, and their adaptation with host plants dated as far back as the Carboniferous period (Krings et al., 2012). It has been speculated that there may be at least 1 million existing endophytic fungal species (Strobel and Daisy, 2003) as each of the 300,000 species of land plants can host at least one or more endophytes (Bacon and White, 2000). Blackwell (2011) has a higher projection of the endophyte diversity of approximately 5.1 million species, based on high-throughput sequencing methods. These estimates on endophyte diversity cover plants from various ecosystems, which include deserts, savannahs, grasslands, temperate forests, Arctic tundra, croplands, mangroves, and tropical forests (Arnold, 2007, 2008; Arnold and Lutzoni, 2007). Of the many ecosystems, findings revealed that endophyte diversity is richest in the tropical regions, as diversity decreases in regions toward the temperate northern boreal forests (Arnold, 2007; Arnold and Lutzoni, 2007). At higher latitudes, fewer species were found compared to endophyte communities in the tropics. Although endophyte communities in the tropics are dominated by only a few classes, the species number is larger (Arnold et al., 2000; Arnold and Lutzoni, 2007). As such, it is generally perceived that endophyte diversity is higher in the tropics than any other region.

Countries with the tropical climate are designated as tropical countries or countries in the tropics. These tropical countries are positioned along the Equator region, within the two Tropic limits; the Tropic of Capricorn (southern hemisphere, 23°26′ (23.5°) S) and the Tropic of Cancer (northern hemisphere, 23°26′ (23.5°) N) (Figure 1). This essentially indicates that all countries originating from the North, Central and South America, the Caribbean, the Central, East and West Africa, and the Southeast Asia, which are within the limits of the two Tropics, are clustered as countries in the tropics. One of the significant features of tropical countries is the rainforests. Although the rainforests constitute only a small portion of the total Earth's land area (only 7%), they harbor rich species diversity (50%–70%) (Strobel, 2006). It has been speculated that the diversity of the microorganisms mirrors the diversity of macroorganisms. As such, there is great anticipation that tropical countries as centers of biodiversity of flora and fauna, are equally rich in species diversity of microorganisms and could contribute to the many possible beneficial applications. This further enhances the strategic value of tropical countries as an invaluable center for exploring and harnessing beneficial microorganisms, particularly for discoveries of lesser-known microorganisms such as endophytes.

To date, existing literature would reveal that most of the studies on endophytes are focused on the Northern Hemisphere and New Zealand. Data on endophytes from host plants in the tropical region are relatively limited (Azevedo et al., 2000). This is unfortunate as tropics are considered the center of diversity, with tropical plants postulated to host a great diversity of endophytes. There is, therefore, immense potential that new endophytes can be discovered from the tropics, along with the new bioactive compounds that may benefit various biotechnological applications. This book will discuss on endophytes from the tropics and their ubiquity and diversity in the various host plants from the tropics. The interaction of endophytes with host plant and other organisms will be briefly highlighted. The roles of endophytes as a reservoir for valuable bioactive compounds, enzymes, and antimicrobial agents are also explored and discussed. This includes their use and applications for

FIGURE 1 The tropical countries positioned along the Equator region, within the two tropic limits; the Tropic of Capricorn (southern hemisphere, 23°26′ (23.5°) S) and the Tropic of Cancer (northern hemisphere, 23°26′ (23.5°) N).

agricultural, industrial, bioremediation, biomedicine, bioenergy and biocatalysis, and for biosynthesis and biotransformation of valuable compounds. A short description of several prominent endophytic species from the tropics is also included. The challenges and prospect of research using endophytes from tropics are presented at the end to provide an overview of the future of endophytic research in the tropics.

Endophytes in the Tropics

1.1 DIVERSITY AND UBIQUITY

Endophytes from the tropical regions were first reported by Petrini and Dreifuss (1981). These endophytes were recovered from several tropical regions, which included the French Guiana, Brazil, and Columbia. Today, endophytes in the tropics are found in the vast green rainforest to the mangrove swamps (Figure 2). Tropical endophytes can be isolated from palm trees (Araceae), fruit trees, and plants of the Bromeliaceae and Orchidaceae families (Azevedo et al., 2000). Their association with host plants is primarily beneficial; as herbivore deterrent, as well as in improving photosynthetic efficiency and water acquisition (Pinto et al., 2000; Arnold et al., 2003; Van Bael et al., 2009). Species of bacteria, fungi, and actinobacteria have been recovered as endophytes from various host plants. The bacterial endophytes have long been thought to originate from epiphytes that may have occasionally found its way into the host tissues. Bacterial endophytes often form commensallistic, mutualistic, or pathogenic association with their host (Griffin and Carson, 2015). For fungal endophytes, they were initially classified into three groups based on their ecological grouping. The groups were the mycorrhizal fungi, the balansiaceous or grass endophytes (species of the genera *Balansia, Ephelis, Epichloe, Neotyphodium*), and the non-balansiaceous endophytes (Schulz and Boyle, 2005). The mycorrhizal endophytes were, however, gradually discussed as a separate group from endophytes, as mycorrhiza behaved differently from endophytes. Mycorrhiza form synchronized plant-fungus interaction for specific nutrient-transfer at specialized interfaces, which were not observed in endophytes (Schulz and Boyle, 2005). As such, fungal endophyte communities are primarily of the balansiaceous and the non-balansiaceous fungal endophytes. The latter is more relevant to the tropics as non-balansiaceous

FIGURE 2 Tropical endophytes can be recovered from plants in the regions of (A) green rainforest to the (B) mangrove swamps.

endophytes are typically members of Ascomycota (and its anamorphic stages), and can be isolated from almost every plant (Petrini, 1986; Stone et al., 2000; Huang et al., 2001).

Endophyte diversity and ubiquity are higher in the tropics than any other environments. Strobel and Strobel (2007) isolated approximately 200 endophytic species from 60 angiosperm host plants found in Peru and Bolivia. This demonstrated the diversity of endophytes in angiosperms. Kruasuwan and Thamchaipenet (2016) isolated bacterial endophytes from roots of sugarcane and revealed the discovery of multiple families (Enterobacteriaceae, Bacillaceae, Micromonosporaceae, Moraxellaceae, Micrococcaceae, Pseudomonadaceae, Paenibacillaceae, Staphylococcaceae, Streptosporangiaceae, Streptomycetaceae, and Thermomonosporaceae). Of these, they found *Streptomyces* sp. was most frequently isolated ($n = 47$, 34.81%) while *Pseudomonas* sp. was the least isolated ($n = 1$, 0.74%). Other genera of bacterial endophytes include *Acinetobacter, Actinomadura, Bacillus, Enterobacter, Kosakonia, Lysinibacillus, Microbispora, Micrococcus, Micromonospora, Pantoea, Paenibacillus,* and *Staphylococcus.* The findings reaffirmed the rich and ubiquitous nature of endophytes (in this case, bacterial endophytes) in the tropics. The associations of endophytes with host plants are primarily beneficial; as herbivore deterrent, as well as in improving photosynthetic efficiency and changes to water relation (Pinto et al., 2000; Arnold et al., 2003; Van Bael et al., 2009).

The higher endophyte diversity observed in the tropical forests is also linked to the greater plant species, abundance of older plants and the relatively stable weather conditions (Hyde and Soytong, 2008). The frequency of endophytic infection is higher in older plants, which are abundant in the tropics especially in the rainforests. The endophyte communities in this region are also relatively stable as tropical environments do not experience prominent prolonged dry periods, which are known to depress endophyte diversity (Suryanarayanan et al., 2011). In fact, high precipitation, which is common in the tropics, has been observed to increase endophytic infection (Rodrigues, 1994; Suryanarayanan et al., 1998; Wilson 2000). Other factors are also known to influence the abundance and ubiquity of endophyte communities in the tropics. Abiotic factors such as desiccation, UV radiation, and relative humidity, are often linked to endophyte distribution and abundance. These factors have a temporal effect on the number of endophyte infections per leaf, their composition, and diversity (Higgins et al., 2014). In the wet tropical forests, persistent humid conditions are speculated to influence endophyte diversity as well (Lodge et al., 1996; Arnold and Herre, 2003). In recent years, several other factors have been suggested as possible influencers of the endophyte communities. They include biogeographic history of host plant and fungi (Arnold and Lutzoni, 2007; Arnold et al., 2009; U'Ren et al., 2012), soil type and land use, and possible endophyte-host-specificity (Suryanarayanan et al., 2002; Pandey et al., 2003; Murali et al., 2007). Nevertheless, there have been postulations that localized spatial structure (e.g., leaf litter density) may influence (to a certain degree), endophyte biodiversity and subsequently subtly impact their biodiversity at the regional and larger scales (Arnold and Herre, 2003; Herre et al., 2007).

Nevertheless, the culminating evidence this far presents greater evidence that tropical endophytes are non-specific, not just toward their host plants, but also toward the forest age and soil conditions. Bacterial OTUs recovered were found to be non-specialized to any particular host plant species (Haruna et al., 2018). As tropical endophytes rarely demonstrate host-specificity, they are said to display host generalism nature (Cannon and Simmons, 2002; Arnold et al., 2003; Murali et al., 2007; Mejia et al., 2008). Host generalism refers to the ability of the tropical endophytes to behave as generalist colonizers, colonizing a variety of host plants in the tropics (May, 1991; Arnold and Lutzoni, 2007). There is possible evidence that host generalism of tropical endophytes exist within levels of host species and their subfamilies (Higgins et al., 2014). The consequence of host-generalism in tropical endophytes is the large number of endophytes isolated from plant tissues. A study by Arnold et al. (2000) using 1992 leaf segments, isolated 1472 isolates, which were further categorized into 418 morphospecies. It is a common occurrence in the tropics, that all leaf tissues are colonized by endophytes (Arnold et al., 2000). The rich diversity, ubiquitous nature, and the host-generalism attributes of endophytes form a complex interactive dynamic to their distribution in the tropics. As such, the estimates of endophyte diversity and ubiquity in the tropics is at best an extrapolative estimate to reflect their species richness, abundance, and dominance (Hawksworth, 1991, 2001).

1.2 ENDOPHYTES AND THEIR TROPICAL HOST PLANTS

The common theory is that endophytes in the tropics are dominated by generalist endophytes (Suryanarayanan et al., 2011; Rajulu et al., 2013). These endophytes can be found in almost any plant species, and have the ability to colonize and exist in different parts of the plant tissues as endophytes, commensals, mutualists or latent pathogens (Stone et al., 2000). There are many different types of plants in the tropics that have been discovered to harbor a wide range of endophyte species. These host plants can be palm trees, fruit trees, woody trees, crop plants, medicinal plants, orchids, bromeliads, as well as mangrove trees or shrubs (Table 1 and Figure 3). One of the most common plants in the tropics that are known to host various endophytes are the tropical palm trees. They are found in the rainforests of Australia, Brazil, Bermuda, China, and in the Southeast Asian regions. The endophytes that reside in these tropical palms are highly abundant and may constitute more than 150 species

TABLE 1 Diversity of endophytes from various host plants found in the tropics

HOST PLANTS	ENDOPHYTE SPECIES	LOCATION, COUNTRY	REFERENCES
Palm trees			
Licuala ramasayi, Euterpe oleracea, Livistona chinensis, Trachycarpus fortune, Licuala palms, Elaeis guineensis	Xylariaceous species, Idriella spp., Aspergillus, Phomopsis, Wardomyces, Penicillium, Glomerella cingulata, Phomopsis, Schizophyllum species	Brazil, Bermuda, Thailand	Rodrigues and Samuels (1990, 1992); Rodrigues et al. (1993); Southcott and Johnson (1997); Taylor et al. (1999); Rungjindamai et al. (2008)
Fruit trees			
banana (Musa acuminata)	Xylaria, Colletotrichum musae, Cordana musae	Brazil	Pereira et al. (1999); Rodrigues and Samuels (1999); Johnson et al. (1992)
hog plum (Spondias mombin)	Guignardia, Phomopsis sp.	Brazil	
mango (Mangifera indica)	Dothiorella sp., Phomopsis mangifera	Asia-Australia	
Woody plants			
Cavendishia pubescens	Phomopsis sp.	olumbia	Bills et al. (1992); Haruna et al. (2018); Murali et al. (2006)
Santiria apiculata A.W. Benn. Var. apiculata (Burseraceae), Microdesmis caseariifolia Planch. Ex Hook. F. (Pandaceae), Rothmannia macrophylla (Hook. F.) Bremek. (Rubiaceae)	Diaporthe sp., Phomopsis sp.	Malaysia, Indonesia, Philippines, Sumatra, Singapore	
Teak (Tectona grandis)	Diaporthe sp., Phomopsis sp.	Southern India	
Crops			
Leguminous plant (Stylosanthes guianensis)	Glomerella cingulata (teleomorph of C. gloeosporioides), Phomopsis sp., Xylaria sp.	South America	Pereira et al. (1993)
Sugarcane	Acetobacter diazotrophicus	Brazil	Reis et al. (1994)

(Continued)

TABLE 1 (Continued) Diversity of endophytes from various host plants found in the tropics

HOST PLANTS	ENDOPHYTE SPECIES	LOCATION, COUNTRY	REFERENCES
Eleusine coracana (finger millet)	Acetobacter diazotrophicus	Tamil Nadu, India	Loganathan et al. (1999)
Caupi bean (Vigna unguiculata)	Streptomyces, nocardiopsis, Streptosporangium, Actinomadura, Nocardia	Brazil	Matsuura (1998)
Red seaweed (Bostrychia radicans)	Phomopsis sp.	Sao Paulo, Brazil	Erbert et al. (2012)
Cactus (Cereus jamacaru)	Cladosporium cladosporioides, Fusarium oxysporum, Acremonium implicatum, Aureobasidium pullulans, Trichoderma viride, Chrysonilia sitophila, and Aspergillus flavus	Caatinga forest, Brazil	Bezerra et al. (2013)
Mangrove plants Euterpe oleracea, Rhizophora apiculata, R. mucronata	Sporormiella minima, Acremonium sp.	Amazon Brazil; India	Rodrigues (1994); Suryanarayanan et al. (1998)
Bromeliads Tillandsia catimbauensis	Penicillium, Talaromyces	Brazilian tropical dry forest (Caatinga)	Silva et al. (2018)
Medicinal plant Brucea javanica (L.) Merr.	Trichoderma sp., Fusarium sp., Penicillium sp., Aspergillus sp.	Indonesia	Nur and Muh (2015)
Curcuma longa L. (turmeric)	Colletotrichum sp.	Indonesia	Bustanussalam et al. (2015); Widowati et al. (2016)

(Continued)

TABLE 1 (Continued) Diversity of endophytes from various host plants found in the tropics

HOST PLANTS	ENDOPHYTE SPECIES	LOCATION, COUNTRY	REFERENCES
Mackinlayaceae	Streptomyces, Micromonospora, Verrucosispora, Actinoplanes, Couchioplanes, Gordonia	Indonesia	Ernawati et al. (2016)
Solanum lycocarpum	Luteococcus, Microlunatus, Streptomyces, Rhodococcus	Brazil	Maitan (1998)
Panaceia (Solanum cernuum Vell.)	Arthrobotrys foliicola, Colletotrichum gloeosporioides, Glomerella acutata, Phoma moricola, Coprunellus radians (Psathyrellaceae)	Brazil	Vieira et al. (2012); Sim et al. (2010); Vieira et al. (2012)
	Yeasts-Candida carpophila, Cryptococcus rajasthanensis, Cryptococcus sp., Kwoniella mangroviensis, Meyerozyma guilliermondii	Brazil	
Leea rubra	Dothideomycetes	Thailand	Chomcheon et al. (2006)
Tylophora indica	Dothideomycetes	India	Kumar et al. (2011)
Tamarind (Tamarindus indica, Linn), Indian mulberry (Morinda citrifolia, Linn)	Endophytic fungi	Thailand	Dalee et al. (2015)
Pereskia bleo, Murraya koenigii, Oldenlandia diffusa, Cymbopogon citratus	Colletotrichum, Phoma, Fusarium oxysporum, Penicillium simplicissimum	Malaysia	Chow and Ting (2015)
Acanthus ilicifolius	Lasiodiplodia sp.	China	Chen et al. (2015)
			(Continued)

TABLE 1 (Continued)   Diversity of endophytes from various host plants found in the tropics

HOST PLANTS	ENDOPHYTE SPECIES	LOCATION, COUNTRY	REFERENCES
Houttuynia cordata Thunb. honeysuckle (Lonicera japonica Thunb.)	Chaetomium globosum Fusarium sp. Eutypa, Oudemansiella, Fusarium, Phomopsis	China China China	Pan et al. (2016); Zhang et al. (2016a); Zhang et al. (2015); Zhang et al. (2016b)
Edgeworthia chrysantha Lindl Dalbergia odorifera	Periconia sp. F 31	China	
Solanum trilobatum	Bacillus subtilis	India	Bhuvaneswari et al. (2015)
Ocimum sanctum (tulsi, Ayurvedamedicine as India's queen of herbs)	Glomerellaceae, Hypocreaceae, Nectriaceae, Xylariaceae, Chaetomiaceae, Diaporthaceae	India	Chowdhary and Kaushik (2015)
Garcinia lancifolia Roxb. (Clusiaceae family)	Bacillus cereus, B. megaterium, Staphylococcus sp., Corynebacterium xerosis, Corynebacterium kutscheri	India	Doley and Jha (2016)
Indian snakeroot (Rauwolfia serpentina Benth)	Colletotrichum gloeosporioides, Penicillium sp., A. awamori	India	Nath et al. (2015)
Solanum xanthocarpum Ocimum sanctum, Aloe vera A. indica, Citrus limon, Gossypium hirsutum, Magnolia champaca, Datura stramonium, Piper betle, Phyllanthus emblica	Aspergillus terrus, Phomopsis vexans Cladosporium sp., Rhizoctonia sp., Aspergillus sp., Chaetomium sp., Biosporus sp., Fuzarium sp., Curvularia sp., Cladosporium sp., and Colletotrichum sp. A. fumigatus, Aspergillus niger, Fusarium solani, Aspergillus repens, Alternaria alternata, Alternaria sp., Phoma hedericola, and F. oxysporum	India India India	Parthasarathy and Sathiyabama (2015); Yadav et al. (2015); Patil et al. (2015)

(Continued)

TABLE 1 (Continued) Diversity of endophytes from various host plants found in the tropics

HOST PLANTS	ENDOPHYTE SPECIES	LOCATION, COUNTRY	REFERENCES
Vismia latifolia (Hypericaceae)	Xylariales (Xylaria cubensis is the most represented)	Amazon Rainforest	Nebel et al. (2001)
Vernonia amygdalina Del., Calotropis procera, Catharanthus roseus L., Euphorbia prostrata Ait., Trigonella foenum-graecum L.	Alternaria, Cladosporium, Phoma, Chaetomium, Drechslera, Curvularia, Bipolaris, Paecilomyces, Emericella, Aspergillus	Sudan	Khiralla et al. (2015)
Ardisia colorata, Molineria latifolia, Zingerberaceae sp., Costus speciosus	Sporothrix sp. (KK29FL1)	Kuala Keniam and Kuala Trenggan within the National Park, Pahang, Malaysia	Hazalin et al. (2009)
Forage grass P. maximum, P. purpureum	Sarocladium sp., Ramichloridium sp., Meira sp. and Sporisorium sp., Acremonium sp., Cladosporium sp., Paraconiothyrium sp.	Brazil	daCosta Maia et al. (2018)
Mangrove shrub tree (Aegiceras corniculatum)	Colletotrichum, Alternaria, Phomopsis, Pestalotiopsis, Guignardia, Cladosporium	Coastal and estuarine areas of Beibu Gulf, China	Gong et al. (2014)
Others tartary buckwheat (Fagopyrum tataricum)	Alternaria, Bionectria, Botryosphaeria, Fusarium, Guignardia, Nectria, Neonectria, Phomopsis, Pseudocercospora, Verticillium spp.	China, India, Bhutan, Nepal	Zhong et al. (2017)

(Continued)

TABLE 1 (Continued) Diversity of endophytes from various host plants found in the tropics

HOST PLANTS	ENDOPHYTE SPECIES	LOCATION, COUNTRY	REFERENCES
Camellia sinensis	*Diaporthe* sp.	Indonesia	Agusta et al. (2006);
Stryphnodendron adstringens	*Phomopsis* sp.	Brazil	Carvalho et al. (2012)
Orchid *Dendrobium* spp. (*D. nobile, D. chrysanthum, D. chrysotoxum*)	*Xylariaceous, Fusarium, Colletotrichum, Phomopsis, Nemania* spp., *Annulohypoxylon* spp., *Nodulisporium* spp.	China	Chen et al. (2013)
Orchids	Endophytic yeasts *Rhodotorula mucilaginosa, Bensingtonia* sp., *Candida parapsilosis*	Brazil	Vaz et al. (2009)

FIGURE 3   Tropical endophytes are harbored by various host in the tropics, from crops such as (A) rice, (B) sugarcane, to fruit trees such as (C) tangerine, (D) cacao, (E) banana, to (F) palm trees.   (*Continued*)

(G)

FIGURE 3 (Continued)   Tropical endophytes are harbored by various host in the tropics, from crops such as (G) orchids.

including mycelia sterilia. Their abundance is estimated at 5.7 fungal species per host plant (Hawksworth, 1991). Woody plants (trees) and non-woody plants can both host endophytes. Examples of non-woody plants with endophytes are *Gynoxis oleifolia* (Compositae) (Fisher et al., 1995), grain crops (Graminae) with diazotrophic bacteria (*Herbaspirillum*) (Olivares et al., 1996), and maize with endophytic actinomycetes (Araujo et al., 1999). Other less common hosts include tropical seaweeds (Thirunavukkarasu et al., 2011), cactus (Bezerra et al., 2013), bromeliads (Silva et al., 2018), and mangrove plants (*Euterpe oleracea*, *Rhizophora apiculata*, and *R. mucronata*) (Rodrigues, 1994; Suryanarayanan et al., 1998).

The tropical fruit trees are also excellent hosts of endophytes. Citrus and tangerine plants in Brazil have been found to host endophytes in almost every part of the plant except for the seeds. Endophytes are also found in banana plants (*Musa acuminata*), in which Xylaria is the most common genus isolated, followed by *Colletotrichum musae* and *Cordana musae* (Pereira et al., 1999). Other fruit trees such as hog plum (*Spondias mombin*) and mango (*Mangifera indica*) were found to host endophytic species of *Guignardia* and *Phomopsis* sp. (Rodrigues and Samuels, 1999), and *Dothiorella* sp. and *Phomopsis mangifera* (Johnson et al., 1992), respectively. Other than fruit trees, other cultivated tropical crops also harbor endophytes. Cashew (*Anacardium occidentale*), cacao (*Theobroma cacao*), leguminous plants, sugarcane, tea, are among the many crops that have been used for isolation of endophytes (Pereira et al., 1993; Reis et al., 1994).

Medicinal plants are also excellent hosts of endophytes. In Brazil, a variety of medicinal plants from the family Solanaceae have been found to host actinobacterial, bacterial, and fungal endophytes. *Solanum lycocarpum* hosts endophytic *Lutecoccus, Microlunatus, Streptomyces,* and *Rhodococcus* (Maitan, 1998), while *Solanum cernuum* Vell. (panaceia) hosts an array of fungal and yeast endophytes (Sim et al., 2010; Vieira et al., 2012). Most of these endophytes have antibacterial and antifungal properties (Vieira et al., 2012) against important pathogens such as *Listeria monocytogenes* and *Agromyces lapidis* (Sim et al., 2010). In the Western Ghats of Southern India, medicinal shrubs were found to host endophytic species that are primarily from Ascomycota (8.6%), Coelomycetes (26.0%), Hyphomycetes (28.0%), Mucoromycotina (0.3%) and sterile forms (4.9%). Among these, *Alternaria, Chaetomium, Fusarium, Colletotrichum, Cladosporium, Penicillium, Phyllosticta,* and *Xylaria* species, were the most frequently isolated (Naik et al., 2008). Another medicinal plant *Leea rubra,* harbors endophytic Dothideomycetes, which produces cytotoxic compounds active against Vero cell (Chomcheon et al., 2006). Dothidiomycetes are also found in the medicinal plant *Tylophora indica,* which reportedly are active against fungal plant pathogens (Kumar et al., 2011).

Among the many tropical host plants, tropical orchids and parasitic plant members of Rafflesiaceae are relatively less studied compared to other host plants. In orchids, the endophytes were typically endophytic yeasts such as *Rhodotorula mucilaginosa, Bensingtonia* sp., and *Candida parapsilosis* (Vaz et al., 2009). Endophytic yeasts are not exclusive to orchids though, as they have been recovered from mangrove plants (Statzell-Tallman et al., 2008), *Solanum cernuum* (Vieira et al., 2012), and crops like rice and sugarcane (Nutaratat et al., 2014). The endophytic yeasts recovered from crops are valuable, as they produce plant growth-promotinghormones such as indole-3-acetic acid (IAA) (by *Hannaella sinensis* DMKU-RP45) and antifungal volatile compounds (by *Torulaspora globosa* DMKU-RP31). On the contrary, endophytes from holoparasitic plants *Rafflesia, Rhizanthes,* and *Sapria* (family of Rafflesiaceae), may not render any benefit to the host plant. Instead, the endophytes recovered from the flower shoots and associated bracts were postulated to be latent parasites. Members of Rafflesiaceae, which produce the world's largest flower, is currently believed to harbor the most reduced endophytes as the endophytes persist in the form of uniseriate filaments throughout the vegetative state (Nikolov et al., 2014).

1.3 FACTORS DRIVING ENDOPHYTE COMMUNITIES IN THE TROPICS

Endophyte communities are influenced by several factors, which may alter the dynamics of interaction between endophytes and their host plant, and subsequently the degree of their colonization and distribution in the various environments. These drivers include geographical origin, soil factors, environmental factors (temperature, humidity, rainfall, UV radiation) and perturbations. Of these, environmental factors and perturbations are considered as primary factors as they influence endophyte communities at a higher degree. For temperature and humidity, the constant warm, moist, and humid conditions in the tropics are postulated to give rise to higher bacterial densities than in the temperate regions (Griffin and Carson, 2015). Rainfall influences host species diversity as high rainfall is linked to higher endophyte colonization (Rodrigues, 1994; Wilson and Carroll, 1994; Suryanarayanan et al., 1998; Suresh et al., 1999). Interestingly, it was observed that although endophytic infection is higher with high rainfall, the leaves tend to be infected by the same species suggesting that only endophytic infection of the same species is accelerated (Suryanarayanan et al., 2002). UV radiation influences endophyte communities primarily in the phyllosphere. In the leaf tissues, pigmented bacteria are predominantly found as the pigments absorb radiation and quench oxygen free radicals, thus allowing colonization of endophytic bacteria in the leaves that are exposed to UV radiation (Corpe and Rheem, 1989; Sundin and Murillo, 1999; Kim and Sundin, 2001; Jacobs et al., 2005; Gunasekera and Sundin, 2006). The influence of environmental factors on endophyte communities is rarely the function of one factor, but a response to the multiple stimuli in the environment.

The other major driver of endophytic communities is perturbation. Both natural and anthropogenic perturbations have been reported to impact endophyte infection and colonization in host plants. Natural perturbations are incidences where natural insect or pathogen infection occurs. For example, mining and galling by leaf mining and galling insects is a form of natural perturbation as this activity can impact the fungal microbiome (Christian et al., 2016). On the other hand, anthropogenic perturbations involve activities that are induced by man. Examples include applications of synthetic fertilizers, pesticides, and fungicides (Pedraza et al., 2009; Fischer and Rodriguez, 2013; Nettles et al., 2016), and the occurrence of acid rain (Helander et al., 1993) and wildfires (Huang et al., 2015). These anthropogenic factors can cause shifts in the endophyte communities resulting in either the decrease in endophyte density, increase in the diversity, or changes to the taxonomic composition. Of the many types of endophytes, perturbations appeared to have the most impact on diazotrophic endophytes. This is because fertilizer application is one of the most common cultural practices that, unfortunately, is also easily abused. The supplementation of high nitrogen fertilizers implicated the population of natural diazotrophs (*Azospirillum, Burkholderia, Gluconacetobacter,* and *Herbaspirillum*) in crops (Fuentes-Ramírez et al., 1999). The interlocking interactions of perturbation factors are therefore complex, and may gradually lead to the absence of some endophytes in certain host plants such as in the case of diazotrophic endophytes in crops.

1.4 SPREAD OF TROPICAL ENDOPHYTES AND THEIR ENDOPHYTE-HOST PLANT ASSOCIATION

Tropical endophytes are just like other endophytes in the world. Their spread from one host plant to another is via the typical mechanisms of vertical or horizontal transmittance. The vertical-transfer of endophytes is the primary transfer mechanism found in grass-endophyte infections (Faeth and Saari, 2012). This also applies to diazotrophic endophytic bacteria (*Acetobacter diazotrophicus, Herbaspirillum* spp., *Azospirillum* sp.), which are typically spread via seeds or by vegetative propagation. On the other hand, the horizontal-transfer of endophytes is more common for endophytes harbored in the leaf tissues (Huang et al., 2008; Higgins et al., 2014). They are transmitted by the insect vectors, especially sap-feeding insects. Horizontal-transfer can also spread via rain and wind (Kaneko and Kaneko, 2004; Herre et al., 2007; Herrera et al., 2009). It has also been postulated that local leaf litter may transfer endophytes horizontally. Host plants are exposed to possible colonization by endophytes originating from a neighboring plant or from dead plant materials in the environment as the endophytes remained in the plants upon death (James and Olivares 1998).

In nature, endophytes generally have a positive association with their host plants. Endophyte-host plant associations have been reported to enhance plant growth, and tolerance to biotic and abiotic stresses. Endophytes influence plant growth by modulating the transfer of nutrients to their plant hosts and/ or by eliciting the host plants to produce plant growth hormones (Khan et al., 2012; Kei et al., 2016). Endophytes with plant growth-promoting properties are desirable isolates for application to enhance growth as well as to protect plants from phytopathogens (Kruasuwan and Thamchaipenet, 2016). This makes them ideal candidates as biofertilizers or biocontrol agents.

The plant growth-promoting attributes by endophytes are primarily due to the production of the various growth-inducing enzymes or hormones, namely the enzyme 1-aminocyclopropane-1-decarboxylate deaminase, the hormone indole-3-acetic acid, and other biomolecules produced by the endophytes that aids in nitrogen fixation, phosphate solubilization, and siderophore production (Kruasuwan and Thamchaipenet, 2016). In response to the hormones and enzymes produced, root and meristem cell elongation, and leaf expansion can be observed (Schmelz et al., 2003; Glick, 2005; Saleem et al., 2007). Phosphate-solubilizing endophytes also contribute to improving nutrient acquisition that leads to better plant growth (Maccheroni and Azevedo, 1998). Endophytic diazotrophs (e.g., *Bacillus*, *Enterobacter*), and actinomycetes (e.g., *Microbispora*, *Streptomyces*) are known phosphate solubilizers, with great potential as biofertilizers of the future. Growth-promotion, and in certain cases the yield, is often observed when compatible endophytic strains are introduced to the plants. Rice and sugarcane are examples of crops, which benefited from the association with diazotrophic bacteria (Govindarajan et al., 2007, 2008). Fungal endophytes elicit similar growth-promoting effects as endophytic bacteria. They are capable of producing indole acetic acid, and in solubilizing phosphorus, as well as transferring scarce soil nutrients (e.g., nitrogen, phosphorus, and sulfur) to plant hosts (Spagnoletti and Lavado, 2017). Dark septate fungal endophytes have a further advantage, having the ability to degrade cellulose, starch, casamino acids, urea, lipids, pectin, and gelatin, which are beneficial substrates for plants (Narisawa, 2017). In some rare cases, fungal endophytes have also been associated with increased nitrogen fixation via molecular signaling. The endophytic fungus *Phomopsis liquidambari*, regulates the H_2O_2/NO-dependent signaling crosstalk, which results in increased nodulation and N_2-fixation in the leguminous roots of peanuts (Xie et al., 2017).

The growth-promoting ability of endophytes is beneficial to crops and this forms the principle of using endophytes as biofertilizers. Most studies on biofertilizer development use only the selected growth-promoting endophyte and the impact of the particular strain on plant growth is assessed. Unfortunately, this approach revealed that the growth-promoting effect is not easily replicated in all crops or in various environmental conditions. This is attributed to the complex relationship between endophytes and other microorganisms, as well as with their host plants. In response, some studies attempted evaluations of mixed strains by replicating the set-up using mixed consortia (through co-inoculation) with varying observations. In one study, the combination of endophytes resulted in improved growth of sugarcane than single-strain inoculation. However, with other combinations, the biomass of roots and shoots were significantly compromised than in single-strain inoculation. These observations form the conclusion that the type of endophytes used and their compatibility to coexist with one another is pertinent in rendering growth-promotion of the host plants. There is also a possible preference for colonization in the various plant tissues, such as demonstrated by diazotrophic endophytes and actinomycetes. They appeared to favor colonization in root tissues but less in stem tissues (Kruasuwan and Thamchaipenet, 2016). Endophyte-endophyte compatibility is, therefore, a key factor in the success of mixed consortia. In trials, compatibility occurs when 10^4 CFU g fresh weight^{-1} of endophytes were found in plant tissues, whereas incompatibility of consortia is concluded when a decrease of 10 times occurs when more than two different endophytes were co-inoculated. The poor compatibility of endophytes is magnified by the reduction of plant biomass, which suggests the absence of plant-growth promotion. To date, the general conclusion is that combining isolates into mixed consortia may have implications, but this is very much species and host-dependent.

In addition to plant-growth promotion, endophytes are also capable of enhancing host tolerance toward biotic and abiotic factors. The presence of endophytes in host plants helps host plants develop better tolerance toward pathogenic fungi (Arnold et al., 2003) and deters grazing by herbivores (Jallow et al., 2004; Herre et al., 2007). In a study conducted on oil palms grown in the tropics, endophytes such as *Hendersonia* GanoEF, *Amphinema* GanoEF2, and *Phlebia* GanoEF3 showed antifungal activities toward the pathogenic *Ganoderma boninense* (Noorhaida and Idris, 2009). Another example has endophytic *Burkholderia cenocepacia* 869T2 suppressing Fusarium wilt disease incidence on banana plants (Ho et al., 2015). *B. cenocepacia* 869T2 produced pyrrolnitrin and pyrroloquinoline quinone, which inhibits the pathogen and also promotes plant growth. Non-pathogenic endophytic *Fusarium* species have also been recovered from wild banana species, and have shown potential for early suppression of Fusarium wilt incidence in commercial banana varieties (Ting, 2014). Endophyte association also improved tolerance to abiotic stress such as drought resistance (Bae et al., 2009). This is observed in the association of *Curvularia protuberata* to its tropical grass host, which conferred high soil temperature tolerance to soils in a geothermal region (Redman et al., 2002). Improved tolerance to abiotic stress enables host plants to develop better resilience and adaptability to new habitats (Strobel and Daisy, 2003; Schulz, 2006; Rodriguez et al., 2008; Aly et al., 2011; Friesen, 2013). The endophyte-host plant association has also benefited the host-plants in adapting to nutrient-deprived soils, and environments with high drought, saline, and extreme pH conditions (Aly et al., 2011). It is evident that endophytes through their association with host-plants are effective and beneficial for the well-being of the plant. As such, endophytes have tremendous potential for application as effective inoculants for crop protection and in supporting sustainable approaches in agriculture.

1.5 ENDOPHYTE RESEARCH IN THE TROPICS

Endophyte research within the tropical regions has grown tremendously, particularly with the advent in discoveries related to bioactive compounds and the tools and techniques assisting the discoveries. This rapid increase in demand for tropical endophytes is related to the great plethora of plants found in the tropics, and their capacity in harboring varies species of endophytes. Endophyte research in Asian countries gained rapid momentum in recent years, especially in Indonesia, Malaysia, Thailand, China, and India. These Asian countries have also been credited with exceptional knowledge of medicinal plants with ethnobotanical properties. However, as plants are neither feasible nor economical for long-term production of bioactive compounds, alternatives using microorganisms are sought. And one such alternative is the potential use of endophytes from the medicinal plants for the up-scale production of valuable compounds. This strategy has exponentially increased research in endophytes in the last few decades. The bioactive compounds are useful for pharmaceutical, agricultural, and industrial applications. And the integration of the current-omics technologies (genomics, proteomics, metabolomics) further assisted explorations of endophytes in the tropics.

In Indonesia, several medicinal plants have been identified for the isolation of endophytes. It has been discovered that fungal endophytes from these medicinal plants have many beneficial properties, i.e., antioxidant, anti-mutagenic, and antibiotic. The Makassar fruit plant (*Brucea javanica* (L.) Merr.) was found to harbor endophytic *Trichoderma* sp., *Fusarium* sp., *Aspergillus* sp., and *Penicillium* sp. (Nur and Muh, 2015). These endophytes produce an array of bioactive compounds (i.e., glycosides) that is rich with antioxidant and anti-mutagenic properties. A common plant, the turmeric plant (*Curcuma longa* L.) reportedly host 44 endophytic isolates with most having significant antioxidant properties (Bustanussalam et al., 2015). Another well-known Indonesian medicinal plant, Mackinlayaceae, harbors endophytic actinobacteria (i.e., *Streptomyces*, *Micromonospora*, *Verrucosispora*, *Actinoplanes*, *Couchioplanes*, and *Gordonia*). These actinobacterial species produced compounds such as anthraquinones and lupinacidins A and B, which has antibiotic and antitumor properties (Ernawati et al., 2016).

In Malaysia and her neighboring country Thailand, numerous medicinal plants with known ethnobotanical values have also been explored for endophyte research. Plants such as tamarind (*Tamarindus indica* L.) and Indian mulberry (*Morinda citrifolia* L.), have been used for isolation of endophytes with antimicrobial activity against *Bacillus cereus*, *Escherichia coli*, *Staphylococcus aureus*, *Salmonella typhi*, and *Candida albicans* (Dalee et al., 2015). Other traditional Malaysian medicinal plants investigated include *Pereskia bleo*, *Murraya koenigii*, *Oldenlandia diffusa*, and *Cymbopogon citratus*. These plants are typically used to treat ailments such as diabetes, hypertension, and cancer, thus are ideal host plants for isolating valuable endophytes that may exhibit similar properties. Chow and Ting (2015) studied these plants and discovered 25 of the 89 endophytes are positive for L-asparaginase production, a valuable anti-tumor agent. These species are from the genera *Colletotrichum*, *Fusarium*, *Phoma*, and *Penicillium* sp., with isolates with the most potential identified as *Fusarium oxysporum* and *Penicillium simplicissimum* (Chow and Ting, 2015). Some of the commonly studied endophytic species in Malaysia are illustrated in Figure 4.

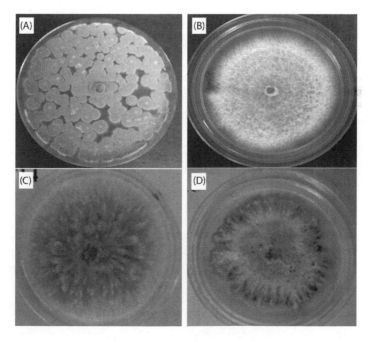

FIGURE 4 Common fungal species studied in Malaysia include (A) *Penicillium*, (B) *Fusarium*, (C) *Diaporthe*, and (D) *Phomopsis*.

In China, different species of medicinal plants are explored for endophyte isolation. Endophytes producing antidiabetic compounds have been isolated from *Acanthus ilicifolius*, a plant with multiple uses for the traditional treatment of coughs, asthmas, hepatitis, lymphadenectasis, and stomach aches. In fact, the antidiabetic compound from the endophyte was discovered to have more potent activity than the common clinical drug acarbose used for treatment (Chen et al., 2015). Endophytes with antimicrobial activities (e.g., *Fusarium* sp.) have also been isolated from honeysuckle (*Lonicera japonica* Thunb.) and *Edgeworthia chrysantha* Lindl (Zhang et al., 2015; Zhang et al., 2016a). *Annona muricata* hosts *Periconia* sp. F 31, which produces compounds with antiviral, cytotoxic, and anti-inflammatory activities (Zhang et al., 2016b). Endophytes from other common Chinese medicinal plants also have the potential for development as biofungicide agent. The medicinal plant *Houttuynia cordata* Thunb., harbors endophytic *Chaetomium globosum* with antifungal activities (100% inhibition) toward important plant pathogens (i.e., *Botrytis cinerea* Pers. ex Fr, *B. cinerea* Persoon, and *Exserohilum turcicum* (Pass.)) (Pan et al., 2016).

In India, most of the research on endophytes is on medicinal plants that are associated with Ayurveda medicine. One of the most common plants used is *Ocimum sanctum* (India's Queen of herbs in Ayurveda medicine) (family Lamiaceae). *O. sanctum* hosts endophytic species of Chaetomiaceae, Diaporthaceae, Glomerellaceae, Hypocreaceae, Nectriaceae, and Xylariaceae. These endophytes have antifungal properties against fungal pathogens *S. sclerotiorum* and *F. oxysporum* with IC_{50} as low as 0.38 mg/mL (Chowdhary and Kaushik, 2015). Another medicinal plant, *Solanum trilobatum* (family Solanaceae), harbors endophytic *B. subtilis*, which produces bacteriocins effective against *E. coli* and *S. aureus*. On its own, *S. trilobatum* produces compounds that are antimicrobial, anti-inflammatory, hepatoprotective, and hemolytic (Doley and Jha, 2016).

It is therefore evident that endophyte research in the tropics is primarily focused on medicinal plants, although other non-medicinal plants are also explored. The selection of medicinal plants for endophyte acquisition appears to be a natural process (although appears bias at times), as it has been postulated that there is a higher probability of acquiring endophytes with valuable compounds from plants sharing similar traits. It now appears that the primary prerequisite for selection is that the host plants must be established beneficial medicinal plants. The plants itself are anticipated to produce bioactive compounds that are antimicrobial, cytotoxic, or antioxidant. Therefore, the endophytes from these host plants are then theorized to express similar attributes as the host plants (Doley and Jha, 2016).

Applications of Endophytes from the Tropics

2

2.1 INTRODUCTION

Literatures would reveal that endophytes have long been recognized for their beneficial roles associated with their host plant. Most of these literature are based on findings in temperate regions. In the early days, endophytes are valued for their role as effective herbivore deterrents (Bills et al., 1992). They then found applications as bioagents to control pathogens and insect pests (Azevedo et al., 2000), and as plant-growth promoters to improve plant growth. The use of endophytes as biofactories of valuable compounds was introduced with the discovery of taxol from the endophytic *Taxomyces andreanae* in 1993 (Stierle et al., 1993). Taxol, or paclitaxel, is a metabolite that has anticancer properties. Taxol was originally extracted from the yew tree *Taxus brevifolia* (Wani et al., 1971). However, with the discovery of taxol production by *Taxomyces andreanea* (Stierle et al., 1993) and later including *Penicillium raistrickii* (Stierle and Stierle, 2000), this has opened doors to the discovery of taxol in many other endophytic species. In addition to taxol, endophytes are now extensively sought for their production of other valuable compounds such as bioactive compounds, antioxidants, and various enzymes and biomolecules with therapeutic properties (e.g., anticancer, antifungal, antidiabetic, immunosuppressant compounds) (Gouda et al., 2016). This highlights the valuable application of endophytes or their bioactive compounds for applications in the agriculture, medicine, and food industry (Shweta and Shaanker, 2010; Kumara et al., 2012).

One of the significant observations on biomolecules from endophytes is their close association with compounds produced by their host plants. For example, compounds such as camptothecin, podophyllotoxin, vinblastine, hypericin, diosgenin, azadirachtin, rohitukine, are similar to bioactive compounds produced by their respective host plants (Shweta and Shaanker, 2010; Kumara et al., 2012). These findings gradually led to the postulation that endophytes could be useful bioresources (as alternatives to plants) for the production of important bioactive compounds (Tan and Zou, 2001). It also demonstrates that the selection of host plant is relatively important in harnessing the attributes desired of a particular endophyte or the compound produced. As such, endophytes from medicinal plants or with ethnobotanical values are often used to isolate endophytes with potent biomolecules or compounds for biomedicine, and for biosynthesis and biotransformation of valuable compounds. To harness endophyte with bioremediation potential, careful selection of tolerant host plants in polluted environments are sought to increase the possibility of isolating endophytes with bioremediation potential. The following chapters will further discuss the application of endophytes from the tropics; in agriculture, biomedicine, industrial, bioenergy, and bioremediation.

2.2 TROPICAL ENDOPHYTES FOR AGRICULTURE

Endophytes are useful in agricultural applications, owing to their ability to suppress disease development as well as enhancing tolerance and adaptations to stressful environments. Thus, endophytes find application as biocontrol agents or biofertilizers, where applications to crops are meant to improve crop tolerance to biotic and abiotic factors and to enhance yield. Endophytic bacteria, fungi as well as actinobacteria, have been reported as effective bioagents. For example, *Alcaligenes piechaudi* and *Kluyvera ascorbata* are effective towards pathogenic *Xanthomonas campestris* pv. *campestris*, the causal agent of cabbage black rot disease (Assis et al., 1998). Fungal endophytes are equally effective biocontrol agents. *Dothidiomycetes* sp. was able to suppress the growth of *Fusarium oxysporum* and *Sclerotinia sclerotiorum* (Kumar et al., 2011), while *Alternaria alternata* inhibited *Plasmopara viticola*

(Musetti et al., 2006). The antifungal activities are attributed to the production of bioactive compounds or metabolites that affects the pathogen. Endophytes, therefore, make ideal biocontrol agents as they can colonize plant tissues asymptomatically and suppress the growth of pathogens by producing antimicrobial compounds (Musetti et al., 2006), compete for nutrient resources and space (Hallmann et al., 1997), and induce host defense mechanisms in the host plants (Berg and Hallmann, 2006).

Biocontrol of insect pests is also possible and has been successful with the use of endophytic entomopathogenic fungi. These endophytes (e.g., species of *Acremonium*, *Beauveria*, *Clonostachys*, and *Paecilomyces*) have bioinsecticidal activities (Vega et al., 2008). When entomopathogenic fungi exist as endophytes in their host plant, they confer host plants with resistance to insect infection. Some endophytic fungi such as *Nodulisporium* sp. from the plant *Bontia daphnoides* produce compounds (nodulisporic acids) that are bioinsecticidal. Nodulisporic acids are effective against the insect pest blowfly (Ondeyka et al., 1997) as well as blood-sucking arthropods (fleas, ticks, bedbugs) (Bills et al., 2012). Another endophyte, *Muscodor vitigenus*, produces the volatile compound naphthalene, which is an effective insect repellent used in mothballs (Daisy et al., 2002).

In addition to tolerance to biotic stress, tolerance to abiotic stress in crops can also be improved with the use of endophytes. Inoculation of endophytes with specific traits such as salt- and temperature-adapted isolates (e.g., *Piriformospora indica*) has helped to improve salt and drought tolerance in *Brassica campestris* (Chinese cabbage) and *Hordeum vulgare* (barley) (Baltruschat et al., 2008; Sun et al., 2010). These plants thrive in salt- and drought-stress conditions as endophytic association has improved plant growth, water retention, and chlorophyll concentration. The tolerance to abiotic stress is attributed to the role of endophytes in inducing antioxidant pathways that could subsequently suppress reactive oxygen species (ROS)-induced oxidative damage to the host plants. This is achieved when endophytes produce antioxidant enzymes, or when endophytes stimulate the plant antioxidant mechanisms to alleviate stress-induced ROS (Hardoim et al., 2015; Egamberdieva and Allah, 2017). These antioxidant mechanisms aid in reducing the impact of ROS on plant membrane dysfunction and hormonal imbalances that otherwise implicates plant health and growth (Egamberdieva and Allah, 2017). Tolerance to abiotic stress, however, is not readily achieved in all plants with the same endophytes. It has been revealed that the response of plants and endophytes to abiotic stress is very much dependent on host species, host genotype, and the environmental conditions (Cheplick, 2004; Rodriguez et al., 2008). Table 2 presents a summary of tropical endophytes, their metabolites, and bioactivities that has beneficial agricultural applications.

TABLE 2 Summary of tropical endophytes, their metabolites, and bioactivities associated with agricultural applications

ENDOPHYTE	HOST PLANT	SECONDARY METABOLITES	BIOACTIVITIES	REFERENCES
Epicoccum nigrum	Sugarcane	Epicorazines A-B, epirodines A-B, flavipin, epicoccines A-D, epipiridones, epicocarines	Epicorazines and flavipin for biocontrol	Brown et al. (1987); Baute et al. (1978); Ikawa et al. (1978); Bamford et al. (1961); Zhang et al. (2007); Wangun and Hertweck (2007); Madrigal et al. (1991); Madrigal and Melgarejo (1995)
Pestalotiopsis microspora	Found in many tropical plant species	Ambuic acid	Antifungal (against Fusarium), anti-oomycete agent (against Pythium), quorum-sensing inhibitor	Li et al. (2001); Fuqua et al. (1994)
Colletotrichum sp. Colletotrichum gloeosporioides	Artemisia annua Amygdalus mongolica	Colletonoic acid antimicrobial	Antibacterial, antifungal, and antialgal activities antifungal against Helminthosporium sativum	Bills et al. (2002); Hussain et al. (2014b); Zou et al. (2000)
Cordyceps dipterigena	Tropical plant species	Cordycepsidone A	Antifungal activity against Gibberella fujikuroi	Varughese et al. (2012)

(Continued)

TABLE 2 (Continued) Summary of tropical endophytes, their metabolites, and bioactivities associated with agricultural applications

ENDOPHYTE	HOST PLANT	SECONDARY METABOLITES	BIOACTIVITIES	REFERENCES
Cryptosporiop-sis quercina	Tropical plant	Cryptocin	Potent activity against Pyricularia oryzae, Fusarium oxysporum, Geotrichum candidum, Rhizoctonia solarum, S. sclerotiorum, Py. ultimum, Phytophthora cinnamon, Ph. Citrophthora	Li et al. (2000)
Pestalotiopsis jesteri	Plant from Papua New Guinea	Jesterone	Anti-oomycete activity	Li and Strobel (2001)
Muscodor crispans	Ananas ananassoides (wild pineapple) in the Bolivian Amazon Basin	Antifungal and antibacterial	Antifungal and antibacterial-Pythium ultimum, Alternaria helianthi, Botrytis cinerea, Fusarium culmorum, F. oxysporum, Phytophtora cinnamomi, Ph. palmivora, Rhizoctonia solani, Sclerotinia sclerotiorum, Verticillium dahliae, Xanthomonas axonopodis pv. citri	Mitchell et al. (2010)
Gliocladium sp.	Eucryphia cordifolia	Annulene	Antimicrobial	Stinson et al. (2003)

(Continued)

TABLE 2 (Continued)   Summary of tropical endophytes, their metabolites, and bioactivities associated with agricultural applications

ENDOPHYTE	HOST PLANT	SECONDARY METABOLITES	BIOACTIVITIES	REFERENCES
Xylaria feejeensis	Plant from Peru and Bolivia	Nonenolide named xyolide	Antifungal towards Pythium ultimum	Baraban et al. (2013)
Lasiodiplodia theobromae JF766989	Piper hispidum	Crude extracts	Antifungal activities towards phytopathogens	Orlandelli et al. (2017)
Phomopsis longicolla, Phomopsis sp. HNY29-2B	Endangered mint Dicerandra frutescens	Antibiotic and cytotoxic dicerandrols A, B, and C	Xanthomonas oryzae, Dicerandrol A is antimicrobial against Gram-positive bacteria and yeasts	Wagenaar and Clardy (2001); Lim et al. (2010); Ding et al. (2013)
Phomopsis sp. co-cultured with Alternaria sp.	Mangrove plants	L-leucyl-4-hydroxyprolyl-D-leucyl-4-hydroxyprolyl	Inhibits crop pathogens (Gaeumannomyces graminis, Rhizoctonia cerealis, Helminthosporium sativum, and Fusarium graminearum)	Li et al. (2014)
Epicoccum purpurascens, Epicoccum nigrum	Recovered from tropical orchids in Brazil	Epicorazine A and B, epicorazine X and Y, and flavipin	Antifungal activity against the pathogenic fungi Pythium intermedium, Phytophthora cinnamomi, Fusarium oxysporum, C. albicans, C. krusei	Brown et al. (1987); Vaz et al. (2009)

(Continued)

TABLE 2 (Continued) Summary of tropical endophytes, their metabolites, and bioactivities associated with agricultural applications

ENDOPHYTE	HOST PLANT	SECONDARY METABOLITES	BIOACTIVITIES	REFERENCES
Pestalotiopsis microspora Pestalotiopsis jesteri	Recovered from Torreya taxifolia Sepik river area of Papua New Guinea	Torreyanic acid jesterone and hydroxy-jesterone	Antifungal compounds pestaloside, pyrones, pestalopyrone, and hycroxypestalopyrone antifungal activity against phytopathogens	Lee et al. (1996); Li and Strobel (2001)
Nigrospora sphaerica	Medicinal plant Smallanthus sonchifolius	8-hydroxy-6-methoxy-3-methylisocoumarin and (22E,24R)-ergosta-4,6,8(14), 22-tetraen-3-one	Reduce spore germination or growth of pathogenic F. cxysporum, Colletotrichum gloeosporioides	Gallo et al. (2009); Kim et al. (2001); Rosa et al. (2012)
Diaporthe miriciae	Copaifera pubiflora and Melocactus ernestii.	Cytochalasins; Cytochalasins H and J	Antifungal; pathogens Coletotrichum fragariae, Coletotrichum gloeosporioides, Coletotrichum acutatum, Botrytis cinerea, Fusarium oxysporum, Phomopsis obscurans, Phomopsis viticola	Carvalho et al. (2018); Chapla et al. (2014); Cimmino et al. (2008); Lin et al. (2009a); Xu et al. (2009)

2.3 TROPICAL ENDOPHYTES FOR BIOMEDICINE AND BIOSYNTHESIS OF VALUABLE COMPOUNDS

Endophytes for biomedicine are endophytes that have the capacity to produce valuable compounds or secondary metabolites that have therapeutic properties. The key step in harnessing valuable endophytes and the compounds they produce is the selection of the host plant. A hypothesis exists that the probability of isolating endophytes with desired pharmacological activities is relatively higher when plants exhibiting similar properties are selected (Strobel, 2002). It has been postulated that endophytes, through coexistence and coevolution with the host plants, may sometimes acquire the ability to synthesize compounds that are similar to the host plants (Tan and Zou, 2001; Strobel and Daisy, 2003). This includes metabolites, compounds and derivatives, and even volatile compounds and essential oils. This explains the preference of sourcing endophytes from medicinal plants or plants with known ethnobotanical properties (Huang et al., 2008; Rosa et al., 2010; Higginbotham et al., 2013).

The use of endophytes as sources for bioactive compounds is now a more attractive option than other microorganisms. Studies have suggested that endophytes appeared to be more metabolically innovative than soil fungi as a result of their constant interaction with their host plants (Schulz et al., 2002). Endophytes, therefore, are capable of producing metabolites that are rather unique (in some cases novel metabolites), as means of adaptation to defense mechanisms of the host plant (Stadler and Keller, 2008; Mitchell et al., 2010). The defense mechanisms by the host plants often lead to the production of various metabolites. These metabolites are non-selective in nature, thus endophytes residing in the host plants learn to overcome or cope with the metabolites by developing enzymatic pathways to convert or degrade the compounds, giving rise to new metabolites (Kaul et al., 2014).

2.3.1 Antimicrobial Compounds

The compounds produced by endophytes can be in the form of volatile or non-volatile forms, and with a variety of antimicrobial, antioxidant, antitumor, anti-inflammatory, and antiallergic properties. Of these, antimicrobial compounds are the most extensively studied for their potential as antibiotics (Matsuura, 1998). Endophytes from host plants in the tropics are reportedly effective producers of antimicrobial compounds. Endophytic actinomycetes, for example, produce antimicrobial compounds effective against *Staphylococcus aureus* and *Bacillus subtilis* (Britto, 1998). Fungal endophytes such as *Phoma moricola* from *Solanum cernuum* and *Garcinia parvifolia*, also produce antimicrobial compounds effective against *Listeria monocytogenes* and *Agromyces lapidis* (Sim et al., 2010). Endophytic yeasts (*Rhodotorula mucilaginosa, Bensingtonia* sp., *Candida parapsilosis*) from tropical orchids were also inhibitory towards common bacterial pathogens (*E. coli, S. aureus, B, cereus, S. typhimurium*). The efficacy of antimicrobial compounds produced by the endophytes has been found to be comparable to existing antibiotics. This was revealed by Vieira et al. (2012) using compounds from endophytic *Coprinellus radians, Cryptococcus rajasthanensis, Colletotrichum gloeosporioides*, and *Fusarium* sp., which demonstrated MIC values similar to values derived from chloramphenicol, when tested against *E. coli* (MIC 2–8 μg/mL) and *S. aureus* (MIC 2–16 μg/mL).

Although endophytic species are abundant in the tropics, not all of these endophytes are natural-producers of antimicrobial compounds. Often in a sampling exercise using 100–150 isolates, the probability of recovering antibiotic-producers is 50%–70% (Smith et al., 2008). Nevertheless, species of *Pseudomonas* and *Phomopsis* have been relatively consistent in yielding positive results for antimicrobial activity. *Pseudomonas* sp., found ubiquitously in almost any host plants in the tropics, produces the antifungal peptides (pseudomycins) (Strobel et al., 2004; Berdy, 2005), which strongly inhibits pathogens of the Dutch elm disease (*Ceratocystis ulmi*) and Black Sigatoka disease of banana (*Mycosphaerella fijiensis*) (Harrison et al., 1991; Ballio et al., 1994). For *Phomopsis* spp., found in tropical marine red seaweed (*Bostrychia radicans*) (Erbert et al., 2012), medicinal plants and crop plants, the antifungal activities of this endophyte are attributed to a variety of compounds produced. This includes phomopsidin (antimicrotubule agent) (Kobayashi et al., 2003), phomoxanthone A (antifungal) (Elsasser et al., 2005), phomodiol (antifungal) (Horn et al., 1995), biaryl ethers (herbicidal) (Dai et al., 2005), phomosines (algicidal) (Krohn et al., 1995), phomalactone derivative (cytokine inhibitors) (Wrigley et al., 1999), and phomopsichalasin (antimicrobial) (Horn et al., 1995).

Recent studies have also highlighted that many of the tropical endophytes produce antimicrobial compounds that are similar to their host plants. For example, the endophytic fungi *Fusarium proliferatum* was found to produce sanguinarine, a compound with antibacterial, anti-inflammatory, and anti-helminthic properties that are also found from the host plant *Macleaya cordata* (plume poppy) (Wang et al., 2014). Another antimicrobial compound, rhein, is produced by endophytic *Fusarium solani*, which is also produced by the host plant *Rheum palmatum* L. (the Chinese rhubarb) (You et al., 2013). Another beneficial compound, chlorogenic acid, is produced by both the host plant *Eucommia ulmoides* Oliver and its endophyte (*Sordariomycete* sp.) (Chen et al., 2010).

The volatile antimicrobial compounds produced by endophytes depends on the species. For example, *Muscudor crispans* produce volatiles that are generally less toxic, but the other *Muscudor* species, *M. albus*, produces toxic volatiles (i.e., azulene or naphthalene derivatives). Volatile compounds that are generally safe compounds are classified by the US Food and Drug Administration as generally regarded as safe (GRAS) compounds (FDA website www.fda.gov/). GRAS compounds find applications in agricultural (mycofumigation), food preservation and in industrial uses (Mitchell et al., 2010). A review by Suryanarayanan et al. (2009) captures the many types of novel metabolites produced by a variety of endophytes. The common antimicrobial compounds produced by tropical endophytes are summarized in Table 3. It is also noted that bacteria endophytes rarely produce the antimicrobial secondary metabolites produced by endophytic fungi. This suggested that fungal endophytes may produce metabolites that are distinct from bacterial endophytes, and a certain degree of exclusivity of active metabolites to fungal endophytes may exist (Haas and Défago, 2005; Lugtenberg and Kamilova, 2009; Pliego et al., 2011).

2.3.2 Anticancer Compounds

One of the very first discoveries of beneficial compounds produced by endophytes are anticancer compounds. The taxol-producing endophytic *Taxomyces andreanae* was found to produce taxol, a compound that is also produced by the host plant, the Yew tree. Since then, other endophytes have been explored as alternatives for the production of useful plant-like compounds (Pu et al., 2013; Xiong et al., 2013; Palem et al., 2015). Over the years, more endophytes in the tropics have been documented to produce valuable anticancer compounds that are similar to their host plants. This includes camptothecin (CPT) (from *Camptotheca acuminata*), and vinblastine and vincristine (from *Catharanthus roseus*), which have been developed as anticancer agents (Pu et al., 2013; Palem et al., 2015). In addition, some endophytes also produce enzymes that have antitumor activities. L-asparaginase, for example, is one such valuable enzyme. L-asparaginase is capable of depleting L-asparagine in the blood, thus depriving tumor cells of acquiring L-asparagine, leading to cell death. L-asparaginase is, therefore, used to treat acute lymphoblastic leukemia (ALL), the most frequent form of cancer among children (Azevedo-Silva et al., 2010). Sarquis et al. (2004) reported the production of L-asparaginases by species of *Aspergillus*, *Penicillium*, and *Fusarium*. Thirunavukkarasu et al. (2011) reported *Alternaria*, *Chaetomium*, *Cladosporium*, *Colletotrichum*, *Curvularia*, *Nigrospora*, *Paecilomyces*, *Phaeotrichoconis*, *Phoma*, and *Pithomyces* as producers of L-asparaginase. Antitumor compounds are also

TABLE 3 Summary of tropical endophytes, their metabolites, and bioactivities associated with antimicrobial applications

ENDOPHYTE	HOST PLANT	SECONDARY METABOLITES	BIOACTIVITIES	REFERENCES
Colletotrichum sp. Colletotrichum gloeosporioides	Artemisia annua Amygdalus mongolica	Colletonoic acid antimicrobial	Antibacterial, antifungal, and antialgal activities against bacteria	Bills et al. (2002); Hussain et al. (2014b); Zou et al. (2000)
Pestalotiopsis microspora	Terminalia morobensis	Pestacin and isopestacin	Antimicrobial	Strobel et al. (2002); Harper et al. (2003)
Phomopsis sp.	(Not specified)	Phomopsicha-lasin	Antibacterial against Bacillus subtilis, Salmonella enterica, Staphylococcus aureus	Horn et al. (1995)
Phomopsis longicolla	Red seaweed Bostrychia radicans	18-deoxycy-tochalasin H, mycophenolic acid, and dicerandrol C	Dicerandrol C is antimicrobial (against Staphylococcus aureus, Staphylococcus saprophyticus)	Erbert et al. (2012)
Muscodor albus	Cinnamomum zeylanicum (cinnamon tree)	28 volatile compounds (alcohols, esters, ketones, acids, and lipids), particularly esters isoamyl acetate	Antimicrobial	Worapong et al. (2001); Strobel et al. (2001)
Gliocladium sp.	Eucryphia cordifolia	Annulene	Antimicrobial	Stinson et al. (2003)

(Continued)

TABLE 3 (Continued)   Summary of tropical endophytes, their metabolites, and bioactivities associated with antimicrobial applications

ENDOPHYTE	HOST PLANT	SECONDARY METABOLITES	BIOACTIVITIES	REFERENCES
Penicillium tropicum	Sapium ellipticum	Cyclohexapeptide, penitropeptide, new polyketide penitropone	Weak antimicrobial and cytotoxic activities	Zeng et al. (2016)
Cylindrocarpon sp.	Tropical plant Sapium ellipticum	Polyketide cylindrocarpones A–C, pyridone alkaloid cylindrocarpyridone, pyrone cylindropyrone, lamellicolic anhydride, 5-chloro-6,8,10-trihydorxy-1-methoxy-3-methyl-9(10H)-anthracenone, 1-O-methylemodin, 5-chloro-1-O-methylemodin, dihydroramulosin, pyrrocidine A, and 19-O-methyl-pyrrocidine B	Cytotoxic activity, antibacterial activity	Kamdem et al. (2018a)
Phoma sp.	Tropical plant	Phomalacton, (3R)-5-hydroxymellein, and emodin	Antifungal, antibacterial, and algicidal properties towards Microbatryum violaceum, Bacillus megaterium, Chlorella fusca	Hussain et al. (2014a)
Diaporthe sp.	Camellia sinensis	Biotransform (+)-catechin (flavan-3-ol) and (−)-epicatechin via oxidation	Derivatives 3,4-cis-dihydroxyflavans	Agusta et al. (2005); Shibuya et al. (2005)

(Continued)

TABLE 3 (Continued)   Summary of tropical endophytes, their metabolites, and bioactivities associated with antimicrobial applications

ENDOPHYTE	HOST PLANT	SECONDARY METABOLITES	BIOACTIVITIES	REFERENCES
Colletotrichum, Alternaria, Phomopsis, Pestalotiopsis, Guignardia, Cladosporium	Aegiceras corniculatum (mangrove shrub/tree)	Crude extracts	Antibacterial activity	Gong et al. (2014)
Alternaria, Bionectria, Botryosphaeria, Fusarium, Guignardia, Nectria, Neonectria, Phomopsis, Pseudocercospora and Verticillium spp.	Fagopyrum tataricum (tartary buckwheat)	Crude extracts	Antimicrobial activity against Bacillus subtilis, Staphylococcus aureus, Agrobacterium tumefaciens, Escherichia coli, Pseudomonas lachrymans Inhibition of spore germination of F. oxysporum f. sp. vasinfectum, F. oxysporum f. sp. cucumerinum	Zhong et al. (2017)
Phomopsis longicolla, Phomopsis sp. HNY29-2B	Endangered mint Dicerandra frutescens	Antibiotic dicerandrols A	Antimicrobial against Gram-positive bacteria and yeasts	Wagenaar and Clardy (2001); Lim et al. (2010); Ding et al. (2013)

(Continued)

TABLE 3 (Continued) Summary of tropical endophytes, their metabolites, and bioactivities associated with antimicrobial applications

ENDOPHYTE	HOST PLANT	SECONDARY METABOLITES	BIOACTIVITIES	REFERENCES
Diaporthe	Pandanus amaryllifolius leaves	Diaportheones A and B	Inhibits Mycobacterium tuberculosis H37Rv	Bungihan et al. (2011)
Alternaria, Chaetomium, Cochliobolus, Diaporthe, Epicoccum, Guignardia, Nigrospora, Pestalotiopsis, Phoma, Phomopsis, Podospora, Preussia, Sporormiella, and Xylaria	Baccharis trimera, a Brazilian medicinal plant in the Brazilian savannah	Crude extracts	Antimicrobial	Vieira et al. (2014)
Epicoccum nigrum	Tropical orchids in Brazil	Epicorazine A and B, epicorazine X and Y, and flavipin	Against C. albicans and Candida krusei	Vaz et al. (2009)
Pestalotiopsis microspora	Recovered from Torreya taxifolia	Pestaloside, pyrones, pestalopyrone, and hydroxypestalopyrone	Antifungal	Lee et al. (1996)

(Continued)

TABLE 3 (Continued) Summary of tropical endophytes, their metabolites, and bioactivities associated with antimicrobial applications

ENDOPHYTE	HOST PLANT	SECONDARY METABOLITES	BIOACTIVITIES	REFERENCES
Preussia pseudominima	*Baccharis trimera*	Diarylcyclopentendione	Antifungal (anthraquinones, a diphenyl ether, culpin, preussin, au-ranticins A and B, preussomerins A–F, the cycloartane triterpene, spirobisnaphthalenes, and thiopyranchromenones)	Du et al. (2012)
Diaporthe miriciae	Medicinal plants *Copaifera pubiflora* and *Melocactus ernestii*, *Camptotheca acuminate*	Cytochalasins; Cytochalasins H and J	Antifungal; pathogens *Colletotrichum fragariae*, *Colletotrichum gloeosporioides*, *Colletotrichum acutatum*, *Botrytis cinerea*, *Fusarium oxysporum*, *Phomopsis obscurans*, and *Phomopsis viticola*; (antimicrobial, antitumor, cytotoxic, and HIV protease inhibitors)	Carvalho et al. (2018); Chapla et al. (2014); Cimmino et al. (2008); Lin et al. (2009b); Xu et al. (2009)
Chaetomium sp.	Marine algae	Chaetoglobosins (cytochalasin analogs); chaetomin, chaetoglobosins, chaetoquadrins, oxaspirodion, chaetospiron, orsellides, and chaetocyclinones	Inhibit actin polymerization	Yahara et al. (1982); Lösgen et al. (2007)

produced by *Mucor* sp. (Huang et al., 2001), particularly *Mucor rouxianus* (Miao et al., 2009). The common antitumor agents produced by tropical endophytes are summarized in Table 4.

2.3.3 Functional Polysaccharides

Endophytes are also known to synthesize functional polysaccharides that have values as biomedicine. Endophytic fungi have been reported to produce or excrete functional polysaccharides that are capable of improving biological-functions and have the potential as nutraceuticals in the pharmaceutical industry (Lösgen et al., 2007). These functional polysaccharides have anti-inflammatory and antiallergic activities, other than their known antitumor and antioxidant activities. Functional polysaccharides are produced naturally in response during the plant-endophyte interactions. The most common naturally occurring polysaccharides are the exopolysaccharides (EPS), which are responsible for the formation of biofilm. The exopolysaccharides (EPS), in the form of biofilm, are produced as a response to abiotic stress, during colonization of plants, and as bioelicitors to stimulate host response to biotic and abiotic stress (Lösgen et al., 2007). These polysaccharides are safe, non-toxic and are biodegradable, thus are ideal for various applications. Some of these polysaccharides (e.g., dextran, xanthan, pullulan, and hyaluronic acid) have been defined and developed for food (thickeners, suspensions stabilizers), pharmaceutical (tablet granulation, coating) and medical (wound healing treatments, eye surgeries) applications (Moscovici, 2015). There are also polysaccharides that are high in antiallergic, antioxidant, and antiproliferative activities. Endophytic fungi *Berkleasmium* sp. Dzf12 (from yellow ginger *Dioscorea zingiberensis*) and *Fusarium solani* SD5 from *Alstonia scholaris* (devil tree), produce polysaccharides with antioxidant activities (Li, 2012b; Mahapatra and Banerjee, 2013). For *Diaporthe* sp. (from the medicinal plant *Piper hispidum* Sw), the exopolysaccharide produced have antiproliferative activities against the human breast carcinoma (MCF-7) and hepatocellular carcinoma (HepG2-C3A) cells (Orlandelli et al., 2017). The exopolysaccharide rhamnogalactan, produced by endophytic *F. solani* SD5 (from *A. scholaris*), has antiallergic, anti-inflammatory, and antiproliferative activities (Mahapatra and Banerjee, 2012). Clearly, endophytes from the tropics have the potential to produce valuable functional polysaccharides for biomedical use.

2.3.4 Other Valuable Compounds/Metabolites

The endophytes have also been found to produce compounds with other therapeutic properties, which are similar to compounds produced by the host plants. This includes compounds or metabolites with antihypertensive,

TABLE 4 Summary of tropical endophytes, their metabolites, and bioactivities associated with antitumor applications

ENDOPHYTE	HOST PLANT	SECONDARY METABOLITES	BIOACTIVITIES	REFERENCES
Dothideomycetes	Leea rubra (medicinal plant)	Pyrone derivatives, dothideopyrones A–D	Cytotoxicity against Vero cell	Chomcheon et al. (2006)
Bionectria sp.	African tropical rain forest plant	Cytotoxic penicolinate A, and two new o-aminobenzoic acid derivatives, bionectriamines A and B	Cytotoxic against ovarian cancer cell line A2780	Kamdem et al. (2018b)
Fusarium oxysporum	Catharanthus roseus leaf	Vincristine (biotransformed from vinblastine)	Anticancer agent	Kumar and Ahmad (2013)
Species of Colletotrichum, Fusarium, Phomaand Penicillium	Medicinal plants (Cymbopogon citratus, Murraya koenigii, Oldenlandia diffusa and Pereskia bleo)	L-asparaginase	Antitumour agent	Chow and Ting (2015)
Phomopsis sp.	Kandelia candel (mangrove); Thai medicinal plant Hydnocarpus anthelminthicus	1,3-dihydro-4-methoxy-7-methyl-3-oxo-5-isobenzofuran-carboxyaldehyde; mycoepoxydiene derivatives, deacetylm ycoepoxydiene	Cytotoxic agents	Lin et al. (2005); Prachya et al. (2007)

(Continued)

TABLE 4 (Continued)   Summary of tropical endophytes, their metabolites, and bioactivities associated with antitumor applications

ENDOPHYTE	HOST PLANT	SECONDARY METABOLITES	BIOACTIVITIES	REFERENCES
Phomopsis sp. HCCB04730 Phomopsis sp. strain B27, Phomopsis sp. strain PSU-MA214	(Not specified) Annona squamosa; and a mangrove plant	Dihydronaphthalenone, phomonaphthalenone A, naphthoquinones anthraquinone derivatives (phomolide C, 1-methoxy-8-hydroxy-9,10-anthraquinone, 1,8-dihydroxy-9,10-anthraquinone or 1,8-DHA, cytosporone C, altiloxin A; (2R,3S)-7-ethyl-1,2,3,4-tetrahydro-2,3,8-trihydroxy-6-methoxy-3-methyl-9,10-anthracenedione, ethyltetrahydroanthraquinone)	Anti-HIV and cytotoxic against breast cancer, antibacterial against S. aureus and MRSA	Yang et al. (2013); Lin et al. 2008a); Klaiklay et al. (2012)
Phomopsis sp. Phomopsis longicolla	Marine mangrove plants from South China Sea Plant from South Korea	Exumolide A Fusaristatin A	Cytotoxic inhibits lung cancer cells LU 65, inhibit Xanthomonas oryzae	Yang et al. (2011); Lim et al. (2010)
Phomopsis glabrae	Pongamia pinnata (family Fabaceae)	Depsipeptide (PM181110)	Anticancer activity (human cancer cell lines, human tumor xenografts)	Venekar et al. (2014)
Diaporthe sp. Phomopsis euphorbiae	Tea plant Camelia sinensis Trewia nudiflora	(−)-epicytoskyrin azaphilones phomoeuphorbins A and C	Cytotoxic	Agusta et al. (2006); Yu et al. (2008)
Fusarium oxysporum, Neonectria macrodidym	Pigeon pea (Cajanus cajan (L.) Millsp.)	Cajaninstilbene acid (CSA, 3-hydroxy-4-prenyl-5-methoxystilbene-2-carboxylic acid)	Cyto-protective effects	Gao et al. (2011)

antidiabetic, antiparasitic, and antimalarial properties. Endophytic *Phomopsis* sp. produces the compound pinoresinol diglucoside with antihypertensive properties. This compound is also produced by the host plant *Eucommia ulmoides*, a small native tree in China (Shi et al., 2012; Yan et al., 2017). Another important compound is lovastatin, which has blood-cholesterol lowering effect. This compound is produced by the endophytes found in the host plant *Solanum xanthocarpum* and by the host plant itself (Parthasarathy and Sathiyabama, 2014). Endophytes from the tropical grass *Paspalum conjugatum* (Poaceae) in Panama have also been reported to produce compounds with antiparasitic activities. The endophytes produce sterigmatocystin, along with the new polyketide integrasone B and secosterigmatocystin. As an antiparasitic agent, sterigmatocystin is effective against *Trypanosoma cruzi*, the causal agent of Chagas disease (Almeida et al., 2014).

Some of the compounds produced by endophytes have high antioxidant activities, which are extremely useful as biomedicine. Antioxidant activities are determined from extracts produced and are often produced in high levels by endophytes isolated from medicinal plants. This has been reported for endophytes isolated from the bulbs of *Fritillaria unibracteata* (Anzi-Beimu in Chinese) (Pan et al., 2017). In recent years, endophytes have also been found to produce volatile compounds similar to essential oils produced by the host plant. A study by Monggoot et al. (2017) discovered that species of *Arthrinium*, *Colletotrichum*, and *Diaporthe* from the host plant *Aquilaria subintegra* (lign aloes), could produce volatile compounds such as β-agarofuran, α-agarofuran, β-dihydro agarofuran, δ-eudesmol, and oxo-agarospirol. These compounds resemble the agarwood oil of the host plant *Aquilaria subintegra*. The discovery of endophyte-producing essential oil presents an alternative to source for essential oils, and with the prospect of upscaling the productivity. Table 5 presents a summary of the endophytes producing compounds useful for biomedicine-related applications.

2.3.5 Biotransformation of Valuable Compounds

In some cases, biotransformation may occur to transform some known compounds into novel compounds with different levels of bioactivities. The biotransformation reactions have been reported in fungal endophytes and is typically regulated by stereo-selective reactions of acylation, epoxidation, deracemization, hydroxylation, nitration, oxidation, reduction, and sulfoxidation (Zikmundova et al., 2002; Borges et al., 2009). This has been observed in the endophytic fungus *Curvularia lunata* which biotransform the antibiotic rifamycin-B (clinically less active) to rifamycin-S that is of higher bioactivity.

TABLE 5 Summary of tropical endophytes, their compounds (or metabolites) for biomedicine-related applications

ENDOPHYTE	HOST PLANT	SECONDARY METABOLITES	BIOACTIVITIES	REFERENCES
Fusarium sp., Nigrospora sp.	Marine alga from India, from trees in India	(Not identified)	Antiplasmodial activities against Plasmodium falciparum	Kaushik et al. (2014)
Phoma herbarum JF766995, Schizophyllum commune JF766994	Piper hispidum Sw	(Not identified)	Antifungal, Proteolytic activity	Orlandelli et al. (2015)
Cochliobolus sp.	Tropical plant Piptadenia adiantoides	Cochlioquinone A and isocochlioquinone A	Display leishmanicidal activity	Campos et al. (2008)
Paecilomyces lilacinus, Penicillium janthinellum, Paecilomyces sp.	Symphonia globulifera	(Not identified)	Antiplasmodial against Plasmodium falciparum (PfINDO)	Ateba et al. (2018)
Xylaria sp.	Leaf of Siparuna sp. (Siparunaceae)	Lactones	Antimalarial drugs	Suryanarayanan et al. (2009); Jimenez-Romero et al. (2008)

C. lunata was able to perform the biotransformation of rifamycin due to the expression of rifamycin oxidase (Banerjee and Scrivastava, 1993; Jobanputra et al., 2003). Another valuable compound that is biotransformed by endophytes is vincristine. Vincristine is a more valuable anticancer drug than vinblastine, and this biotransformation is regulated by *Fusarium oxysporum* (Kumar and Ahmad, 2013). Endophytes in the host plant *Aphelandra tetragona* (from South America) have also been found to biotransform phytoanticipin molecules from the host plant, into new active metabolites via various mechanisms (i.e., acylation, hydrolysis, nitration, oxidation, and reduction) (Zikmundova et al., 2002).

In some cases, the endophytes with biotransformation activities could also be used for the purposeful conversion of molecules or metabolites into subset products that are similar to metabolites produced by mammalian cells. This has been reported for some endophytic species of *Phomopsis*, *Glomerella*, *Diaporthe*, and *Aspergillus* spp., where the isolates were capable of biotransforming thioridazine (a neuroleptic drug) to products produced by the mammalian cells (Borges et al., 2008). Similarly, the endophyte *Pestalotiopsis guepini* is capable of biotransforming norfloxacin (a fluoroquinolone antimicrobial agent) into subset products produced by mammalian cells (Parshikov et al., 2001). The advantage of biotransforming molecules into similar products as mammalian cells is that the risks of cytotoxicity are significantly lowered. This makes endophytes highly desirable for use as natural biotransforming agents to produce biomolecules that are less toxic to mammalian cells, which is a key feature in drug discovery. With biotransformation, lead molecules can be transformed or modified into less toxic variants, while maintaining the integrity of the bioactivity of the lead molecules.

Biotransformation is common in endophytes as endophytes would naturally be involved in the detoxification of defense biochemicals produced by their host plants (Zikmundova et al., 2002; Shibuya et al., 2005; Borges et al., 2009). Interestingly, this ability is prominently demonstrated by endophytic latent pathogens (Suryanarayanan and Murali, 2006). Pedras et al. (2005) discovered that endophytic latent pathogens express an array of plant metabolite-detoxifying enzymes to detoxify phytoalexins, which are produced by host plants in response to biotic stresses. As such, their biocatalytic repertoire is broad-spectrum with any of the following mechanisms capable of biotransformation; oxidation, reduction, transamination, and hydrolysis. An example of successful biotransformation using tropical endophytes (including latent pathogens) is demonstrated by endophytic *Alternaria alternata* and pathogenic *Gibberella fujikuroi*. Both isolates were able to biotransform the lignin anticancer agent podophyllotoxin to a novel molecule 4-(2,3,5,6-tetramethylpyrazine-1)-4-demethyl-epipodophyllotoxin, with the latter having lower toxicity to humans compared to the former (Tang et al., 2011). It is therefore evident that tropical endophytic isolates have the ability to biotransform secondary metabolites from their host plants into useful drug agents (Agusta et al., 2005; Shibuya et al., 2005). These evidence established the potential of endophytes as promising biocatalysts and agents in biotransforming natural products.

2.3.6 Biosynthesis of Nanoparticles by Tropical Endophytes

Endophytic research has also made some significant contributions to the most recent technology, nanotechnology. Endophytes have a role in nanotechnology as a substitute for chemical reagents, responsible for synthesizing metal nanoparticles (NPs). Endophytes produce extracts that are rich in secondary metabolites, which are effective in reducing and capping agents when reacted with metal solutions (Golinska et al., 2016). These metabolites, and in some cases with enzymes such as reductases, interacts with metal ions to form desired metal nanoparticles (González-Rodríguez et al., 2012). In fact, extracts from tropical endophytes are preferred as they are associated with medicinal plants, and are known to produce secondary metabolites that are highly bioactive, similar to the rich diverse metabolites of the host plants. As such, endophytes are preferred compared to the use of other common microbes such as soil-inhabiting microbes.

The biogenically synthesized nanoparticles typically share similar attributes with the nanoparticles synthesized conventionally. For example, silver nanoparticles (AgNPs) synthesized by endophytic *Guignardia mangiferae* (isolated from the leaves of medicinal plants) were of 5–30-nm sizes and have spherical-shapes (Balakumaran et al., 2015). These are similar to the characteristics of nanoparticles synthesized with chemical reagents. To date, there have been no reports on the inferiority of endophyte-synthesized nanoparticles compared to conventionally synthesized nanoparticles, in terms of bioactivity. Of the many metal nanoparticles, biogenic syntheses are mostly centered on silver nanoparticles (AgNPs). AgNPs have broad-spectrum antimicrobial activity, effective against both gram-negative and gram-positive bacteria (Raheman et al., 2011; Golinska et al., 2016), as well as having antifungal activities against plant pathogens (Elmoslamy et al., 2017). In some cases, the antimicrobial efficacy of AgNPs is enhanced when used together with antibiotics. This synergistic relation was observed when AgNPs (synthesized by endophytic *Pestalotia* sp.) are applied with the antibiotics gentamycin and sulphamethizole. These combinations of treatments were successful in inhibiting *Staphylococcus aureus* and *Salmonella typhi* (Raheman et al., 2011). This shows the prospect of synthesizing metal nanoparticles with endophytic extracts and their potential for synergistic application with available antibiotics, as a strategy in developing effective antimicrobial agents.

Endophytes, therefore, present a green and more environmental-friendly approach to synthesize valuable metal nanoparticles. It reduces the use of chemical reagents for the synthesis of nanoparticles (NPs), limiting cost and toxicity impact to the environment (Golinska et al., 2016). The synthesized metal NPs find applications in biomedical (nanosilver coatings, implants, treatment of burn and wounds) and pharmaceutical industries (water disinfectant) (Chen and Schluesener, 2008; Zhao et al., 2013).

2.4 TROPICAL ENDOPHYTES FOR INDUSTRIAL USE

2.4.1 Enzymes from Tropical Endophytes

Endophytes for industrial use are typically endophytes that produce enzymes or pigments/colorants that are useful for various biotechnological applications. For enzymes, common enzymes such as amylases, amyloglucosidase, laccase, cellulase, lipase, chitinase, and protease, are often harnessed from endophytes (Gupta et al., 2002; Thirunavukkarasu et al., 2011; Suryanarayanan et al., 2012; Toghueo et al., 2017; Bengyella et al., 2019). Endophytes typically have enzyme-producing capacity as these enzymes are produced to degrade cell-wall polymers of plant cells (Suryanarayanan et al., 2012). This aids in their colonization as well as subsequent role as decomposers of plants. Endophytes from the tropics are an attractive pool of resources for enzymes with adaptability to a range of conditions. This is mainly attributed to the fact that endophytes often are capable of exhibiting similar traits of their host plants growing in a given habitat. Plants in the tropics exist in many challenging habitats, and the endophytes from these plants are expected to be capable of demonstrating some of these tolerances to unfavorable conditions. For example, the discovery of fungal endophytes such as *Colletotrichum* sp., *Phoma* sp., *Phomopsis* sp., *Paecilomyces* sp. and *Phyllosticta* sp. from mangrove plants. Mangrove plants are rich with tannin, and these endophytes have adapted to the tannin-rich environment in the leaves by producing tannase (tannin acylhydrolase). Production of tannase can be easily induced by tannic acid in cultures independent of host, and the tannase enzymes can be extracted and used as a clarifying agent in the food industry to manufacture wine, fruit juice, and instant tea (Kumaresan et al., 2002; Ramírez-Coronel et al., 2003).

Reports have also shown that a variety of enzyme-producing endophytes can be recovered from various tropical plants such as *Cananga odorata* (Lam.) Hook. f. and Thomson, *Terminalia catappa* L., *Terminalia mantaly* H. Perrier, and medicinal plant *Alpinia calcarata* (Haw.) Roscoe (Burkill, 1985; Orwa et al., 2009; Corrêa et al., 2014). The endophytes recovered from these plants include species of *Fusarium*, *Botryosphaeria*, *Chaetomium*, *Cercospora*, *Corynespora*, *Cladosporium*, *Colletotrichum*, *Cylindrocephalum*, *Diaporthe*, *Guignardia*, *Fusicoccum*, *Lasiodiplodia*, *Mycosphaerella*, *Nigrospora*, *Phomopsis*, *Penicillium*, *Pestalotiopsis*, *Phoma*, *Paraconiothyrium*, *Septoria*, and *Xylaria*. The preliminary screening revealed that endophytic isolates are primarily producers of amylase (78.2% positive isolates), followed by cellulase (64.4%), lipase (59.8%), and laccase (59.8%). In most cases, endophytic strains are capable of producing more than a single type of enzyme (Toghueo et al., 2017). In addition, endophytic bacteria from tropical plants are equally capable of producing a variety of beneficial enzymes. Endophytic bacteria from *Rhizophora mangle* (red mangrove) and *Avicennia nitida* (white mangrove) in São Paulo, Brazil, produce amylase, endoglucanase, esterase, lipase, and protease. The bacterial endophytes with high endoglucanase activity were *Brucella* (MBR2.39), *Curtobacterium* (MBR2.21), *Microbacterium* (MCA2.54), *Ochrobactrum* (MBR2.46), and *Stenotrophomonas* (MBR2.29). *Erwinia* sp. (MBA2.19) demonstrated positive lipase activity, while *Curtobacterium* sp. (MBR2.20) was identified for protease production (Castro et al., 2014).

One of the notable findings was the production of protease by some fungal endophyte species. Protease is an enzyme that is typically produced by endophytic bacteria but was detected in *Schizophyllum commune* JF766994 from *Piper hispidum*. In fact, *P. hispidum* hosts approximately 28.57% of endophytes with proteolytic activities (Orlandelli et al., 2015). Other protease-producers include *Curvularia*, *Corynespora*, *Colletotrichum*, *Robillarda*, *Nodulisporium*, and xylariaceous fungi, although bacteria are the more well-known producers of protease (Gupta et al., 2002). Protease from tropical endophytes can be used for the manufacturing of feed, as well as for food (baking, brewing, seasoning, fermented food) and medical applications (denture cleansers) (Rao et al., 1998; Gupta et al., 2002).

There are also endophytes that are capable of producing rather distinct enzymes. For example, endophytic fungi *Curvularia kusanoi* produces the highly desirable laccase enzyme, which has been characterized as pure laccase (EC 1.10.3.2; Mwt: 53.5–64.4 kDa). This enzyme is useful as it is stable at pH8.2 and at 40°C (Vázquez et al., 2018). Endophytic *Aspergillus niger* DR02, *Alternaria* sp. DR45, *Annulohypoxylon stigyum* DR47, *Trichoderma atroviride* DR17 and DR19, and *Talaromyces wortmannii* DR49, are efficient producers of hemicellulases, which are useful for the degradation of lignocellulosic components in biomass. Endophytic *Fusarium oxysporum* PTM7 from leaves of *Croton oblongifolius* Roxb. (Plao yai), are able to produce lipases (Correa et al., 2014). Another species of *Fusarium* (*Fusarium* sp. CPCC 480097), isolated from chrysanthemum stems, was found to produce the novel fibrinolytic enzyme that may have vast potential in thrombolytic therapy. Another uncommon enzyme is chitinase. Chitinase production was detected in some endophytes (*Phomopsis*, *Colletotrichum*, and *Fusarium*), but absent in others (*Alternaria*, *Nigrospora*, *Pestalotiopsis*, and *Phyllosticta*) (Rajulu et al., 2011). Endophytes from the tropics are clearly capable of producing a plethora of enzymes that are beneficial for various applications.

2.4.2 Pigments (or Colorants) from Tropical Endophytes

Like most fungi, endophytes produce pigments as a natural response to protect their biological structures. These pigments are often in the form of polyketide pigments, which can be used as food colorants or for industrial application. Of the many endophytic species, *Monascus* sp. is one of the first few that was extensively studied for pigment production. *Monascus* sp. produces the red and the yellow polyketide pigments (monascus pigments, MPs) that was traditionally regarded as non-toxic, thus it is safe for use as food colorants. One of the early applications of pigments from *Monascus* is in the form of pigment extracts or as dried fermented red rice powder, in which the latter has been used for more than 2000 years in the South East Asian regions (Feng et al., 2012). Nevertheless, despite the widespread use of *Monascus* pigments in the tropics here, these pigments are not approved for use as food colorants by the authorities in the European Union (EU) and the United States (US). The prohibition is attributed to the possible risk of toxin contamination, as the pigments are produced through the secondary metabolite pathway, along with other metabolites that are potential health hazards such as citrinin. Citrinin is highly toxic and has nephrotoxic and hepatotoxic properties.

Other fungal endophytes produce a type of pigment classified as fungal hydroxyanthraquinoid (HAQN) pigments, which are also used as food colorants (Dufosse et al., 2014). In nature, filamentous fungi, lichens, plants, and insects can produce HAQN pigments. Several strains of *Aspergillus* sp. (*A. glaucus*, *A. cristatus*, and *A. repens*) and *Penicillium* sp. (*P. citrinum*, *P. islandicum*) are known to be producers of yellow (emodin, physcion, questin) and red (erythroglaucin, catenarin, rubrocristin) HAQN pigments (Caro et al., 2012). These pigments, however, were also found to have coproduction of mycotoxins such as aspergiolide A, citrinin, cyclochlorotine, erythroskyrin rugulosin, islanditoxin, luteoskyrin, oxaline, secalonic acid D, or tanzawaic acid A. These mycotoxins are naphthoquinones by chemical nature and some may be pigmented. Similarly, pigment production (physcion, erthroglaucin) in species of *Eurotium* (*E. amstelodami*, *E. chevalieri*, and *E. herbariorum*), are accompanied by mycotoxin echinulin (Gessler et al., 2013). The same risks were present in *Fusarium oxysporum* where toxins (i.e., fusaric acid, nectria-furone, monoliformin, and gibepyrones) are coproduced with the pigments (red HAQN pigments).

As such, one of the key criteria for pigment production by tropical endo-phytes and their applications is the non-mycotoxic nature of the pigments and the absence of coproduction of mycotoxins (Dufosse et al., 2014). To address this, endophytic species that produce non-toxic pigments and with-out mycotoxins, are explored. One of the key findings is the discovery of *Talaromyces* species (or formerly known as *Penicillium* sp.), which produce polyketides similar to those produced by *Monascus* sp. The endophytic *Talaromyces aculeatus*, *T. funiculosus*, *T. pinophilus*, and *T. purpurogenus*, produce the polyketide azaphilone (MPA) pigments without the coproduc-tion of citrinin (Mapari et al., 2005, 2009b). This presents *Talaromyces* as possible producers of non-toxic polyketide pigments rather than *Monascus* sp. Another endophytic isolate, *Epicoccum nigrum*, is also reportedly a non-mycotoxigenic pigment-producer (Mapari et al., 2009b). *E. nigrum* produces the polyketide pigment orevactaene, which also has antioxidant properties (Mapari et al., 2008). In addition to non-toxic characteristics, the red and yellow pigments produced by *T. aculeatus* and *E. nigrum*, respectively, also has better photostability when compared to the commercially available red colorant (from *Monascus* sp.) and the yellow colorant (from turmeric) (Mapari et al., 2009a).

There are also endophytes, which produce similar pigments as the host plants. One such endophyte is *Gibberella moniliformis*. This isolate was recov-ered from the leaves of *Lawsonia inermis* L. ("henna" tree), a shrub native to tropical and subtropical regions of Africa, northern Australia, and south-ern Asia (Sharma et al., 2011). This isolate was found to produce Lawsone (2-hydroxy-1, 4-naphthoquinone), or also known as hennotannic acid. This pigment is orange-red in color, and is a popularly used as skin and hair colorant (Sarang et al., 2017). The lawsone produced by endophyte and from the "henna" tree were analyzed spectrometrically using HPLC and ESI–MS/MS analysis, and the pigments were found to be identical. As such, this Sarang et al. (2017) is the first to report of lawsone production by an endophytic fun-gus, independent of the host tissue. More importantly, this revealed that tropi-cal endophytes could produce pigments similar to their host plants, suggesting the possible use of endophytes to mass-produce pigments independent of the host plant. A summary of the pigments produced by endophytic isolates is listed in Table 6.

TABLE 6   Summary of pigments produced by tropical endophytes

PIGMENTS	ENDOPHYTIC ISOLATES	REFERENCES
Melanin, sporopollenin	Most filamentous fungi	Dufosse et al. (2014)
Yellow pigments physcion, macrosporin	*Alternaria* sp. (from marine mangrove tree *Aegiceras corniculatum* in Zhanjiang, Guangdong, South China Sea)	Huang et al. (2011)
Orange questin, yellow asperflavin, brown 2-O-methyleurotinone	*Eurotium rubrum* (from marine mangrove plant *Hibiscus tiliaceus* near Hainan Island, China)	Lugtenberg et al. (2016)
Yellow oils citromycetin, 2,3-dihydrocitro-mycetin	Marine derived *Penicillium bilaii*	Capon et al. (2007)
Tetrahydroauroglaucin (yellow), isodihydroauroglaucin (orange)	*Eurotium* sp. (from mangrove plant *Porteresia coarctata*)	Dnyaneshwar et al. (2002)
Unidentified and unusual blue pigment	*Periconia* sp. (from hypersaline environment, Puerto Rico)	Cantrell et al. (2006)
Novel yellow 2,3-dihydrocitromycetin	*Penicillium bilaii* (from the Australian Huon estuary)	Capon et al. (2007)
Brown bisdihydroanthracenone derivative, eurorubrin or the new orange anthraquinone glycoside [3-O-(a-D-ribofur-anosyl)-questin]	*Eurotium rubrum* (from mangrove plant *H. tiliaceus* around Hainan Island, China)	Li et al. (2009)
Yellow compound dimethoxy-1-methyl-2-(3-oxo-butyl) anthrakunthone	*Fusarium* sp. ZZF60 (from mangrove area, South China Sea)	Huang et al. (2010)
Red alterporriols-K, L, and M	*Alternaria* sp. (from mangrove shrub *Aegiceras corniculatum*, Zhanjiang Guangdong, South China Sea)	Huang et al. (2011)
Pale yellow oil characterized as 1-O-(2,4-dihy-droxy-6 methylbenzoyl)-glycerol	*Penicillium commune* G2M (from mangrove plant *H. tiliaceus*, Hainan Island, China)	Yan et al. (2010)
Red pigment penicillenone	*Penicillium* sp. JP-1 (from *Aegiceras corniculatum* tree, Fujian, China)	Lin et al. (2008b)

(Continued)

TABLE 6 (Continued)   Summary of pigments produced by tropical endophytes

PIGMENTS	ENDOPHYTIC ISOLATES	REFERENCES
Red pigments	*Penicillium purpurogenum* strain SX01 (from *Ginkgo biloba* L.)	Qiu et al. (2010)
Non-carotenoid pigments- purpurogenone, mitorubrin and mitorubrinol (food colourants)	*Penicillium purpurogenum*	King et al. (1970)
Lawsone (2-hydroxy-1, 4-napthoquinone) or also known as hennotannic acid (orange red dye)	*Gibberella moniliformis* (from *Lawsonia inermis*)	Sarang et al. (2017)

2.5 TROPICAL ENDOPHYTES FOR BIOENERGY AND BIOCATALYSIS

The application of endophytes in the generation of bioenergy is relatively new. The discovery of their potential in generating biofuel arises from the fact that endophytic fungi are capable of producing volatile organic compounds (VOCs). Some of these VOCs are rich in hydrocarbons and other oxygenated compounds, and are extremely useful as mycodiesel, an alternative to fossil fuels (Strobel et al., 2008; Wu et al., 2016). Of the many species of endophytic fungus, *Gliocladium roseum* NRRL 50072 (gradually reassigned as *Ascocoryne* sp.) has demonstrated the most potential as a mycodiesel producer. This isolate, when cultured independent of host on oatmeal-based agar and cellulose-based media, produced a variety of volatile hydrocarbons and their derivatives (Strobel et al., 2008; Griffin et al., 2010). VOCs are also produced by endophytic *Nigrograna mackinnonii*, which are rich in terpenes and odd chain polyenesseveral, which are also possible sources for biofuel production. In addition to hydrocarbon-rich VOCs, several other endophytic species produce high concentrations of lipid. *Penicillium brasillianum*, *Penicillium griseoroseum*, *Penicillium* sp. (PAOE), *Trichoderma* sp. and *Xylaria* sp. are known to produce high lipid concentrations that are sources of biofuel precursors (Santosfo et al., 2011).

Another group of endophytes that are potentially able to produce mycodiesel is endophytes that generate energy from the bioconversion of wastes. This is typically the function of endophytic fungi capable of producing lignocellulolytic enzymes (under microaerophilic conditions) that are responsible for the breakdown, degradation, and conversion of cellulosic materials into potential mycodiesel (Suryanarayanan et al., 2012). This has been reported for endophytic *Hypoxylon* sp., *Gliocladium* sp., *Daldinia eschscholzii, and Xylaria* sp. These endophytes deconstruct carbohydrate in biomass, converting lignocellulose into advanced biofuels (Wu et al., 2017). The endophytic *Gliocladium* sp., for example, was able to degrade plant cellulose for the synthesis of complex hydrocarbons, giving rise to hydrocarbons ranging from C_6 to C_{19} (i.e., hexane, heptane, benzene). The hydrocarbons produced were directly synthesized from the cellulosic biomass and in the absence of hydrolytic pretreatments (Ahamed and Ahring, 2011). This, therefore, suggests the many benefits of endophytes in producing mycodiesel, and their potential for use in bioenergy.

2.6 TROPICAL ENDOPHYTES FOR BIOREMEDIATION

Endophytes for bioremediation refer to endophytes that are capable of removing or degrading xenobiotic compounds/pollutants to give rise to either simpler non-hazardous compounds or to generate bioenergy/ biofuel. Endophytes demonstrate this potential when applied independent of their host plant, or when in association with their host plant (typically with a hyperaccumulator plant), or when expressed as a saprophytic decomposer from endophytic stage in plant residues. Endophytes have shown activities in removing metals, dyes, hydrocarbons, other recalcitrant pollutants (e.g., 2,4,6-trinitrotoluene (TNT), trichloroethylene (TCE), polyester polyurethane, phenanthrene), as well as generating bioenergy from the decomposition and breakdown of lignocellulosic constituents in plant biomass.

2.6.1 Metal Bioremediation

The tropical regions are often affected by metal pollutants, which can be found present in both soil and aquatic environments. Early studies would reveal that metal removal potential in endophytes was associated with endophytes that were found in hyperaccumulator plants. The hyperaccumulator-associated endophytes develop metal resistance due to prolonged exposure to high metal concentrations in the environment (Idris et al., 2004). As such, hyperaccumulator-associated endophytes are sourced from hyperaccumulating plants.

There is also further evidence that the degree of heavy metal pollution and the species of plant hyperaccumulator can influence the endophyte diversity (Wei et al., 2014, 2015). For example, *Typha latifolia* (L.) (cattail, family Typhaceae), in India, a known lead (Pb)-phytoaccumulator and pollution indicator, harbors endophytic fungi belonging to the genera *Aspergillus, Fusarium, Penicillium, Phoma,* and *Myrothecium.* These endophytes are able to tolerate Pb. Another example is *Phragmites* sp., a known phytoremediator plant for phytoremediation in sewage or leachate treatment systems. Several fungal species from *Phragmites (Trichoderma, Penicillium, Diaporthe)* have shown metaltolerant characteristics and have been explored for metal removal independent of the host plant (Sim et al., 2016, 2018). Other hyperaccumulator plants include *Portulaca oleracea* (with endophytic *Lasiodiplodia* sp.) (Deng et al., 2014), and grass species (*Festuca arundinacea* and *Festuca pratensis*) with *Neotyphodium* spp. endophytes (Soleimani et al., 2010). Endophyte-assisted phytoremediation as well as the endophytes residing in these plants are therefore an attractive bioremediation approach as they can enhance the phytoremediation process (Lebeau et al., 2008; Gerhardt et al., 2009; Germaine et al., 2009; Rajkumar et al., 2010; Zhang et al., 2011; Weyens et al., 2015).

The majority of studies on endophytes for metal removal are often associated with their phytoremediator host plant. Endophytes in these plants and environment may have adapted to the high concentrations of the metals, thus are more likely to express tolerance and higher efficacy in metal removal. Barzanti et al. (2007) discovered that endophytic bacteria from the Ni-hyperaccumulator plant *Alyssum bertolonii* have high resistance to heavy metal, which assisted plant growth and enhanced Ni hyperaccumulation. Endophytes in other hyperaccumulators, *Solanum nigrum* (Cd-hyperaccumulator) and *Phytolacca acinosa* (Mn-hyperaccumulators), also aid in metal uptake. Similarly, Cd-tolerant endophytes in *Brassica napus* L. increased biomass of the host plant, attributed to enhanced translocation of Cd by *B. napus.* The endophytes aided the host plant in extracting Cd from the Cd+Pb-contaminated soils, which benefited the host plant. The presence of *Neotyphodium* endophytes in grass species (*Festuca arundinacea, F. pratensis*) also facilitated better Cd accumulation in plant tissues compared to endophyte-free grasses in Cd-contaminated soils. As a result, endophyte-infected grasses show higher biomass production than endophyte-free grasses (Soleimani et al., 2010). Interestingly, it has been observed that metal tolerance traits in endophytes differed from plant species or even from the various tissues sampled from the same plant species. This suggested that endophytes from the same host plant may possess varying levels of metal tolerance, due to adaptation of the tissues to the different metal concentrations (Li et al., 2011).

It is clearly evident that due to their adaptations to high metal concentrations, endophytes are not only allowing plants to tolerate metal stress, but to thrive in conditions with high metal concentrations (Deng et al., 2014;

Khan et al., 2017). There are several mechanisms to adaptations or mechanisms expressed to overcoming metal stress. Metals can be easily biotransformed and/or bioaccumulated in the hypha of endophytes, declining their bioavailability and toxicity to host plants (Khan et al., 2017). Endophytes also produce coping mechanisms such as the production of indole-3-acetic acid (IAA). The production of IAA promotes plant growth, leading to better metal bioaccumulation and improving metal removal by plants. Aside from the production of growth hormones like IAA, endophyte association with plants improves metal phytoextraction as well, by modifying the bioavailability, solubility, and transport of metals from soils to plants. This can be achieved through the production of organic acids, chelators, siderophores, as well as performing redox changes. Among the many, siderophore production is the most common in endophytes. Siderophores are low-molecular-mass iron chelators that are able to form complexation with iron. Siderophores are therefore solubilizing agents for iron, extracting iron from minerals or organic compounds, increasing soluble metal concentrations, so that iron uptake is possible under low nutrient conditions. With the presence of siderophore-producing endophytes, uptake of iron and plant growth is still possible under nutrient-limited conditions. In addition to iron, siderophores also form stable complexes with other important metal cations that are essential for plant growth. These include Al, Cd, Cu, Pb, and Zn (Rajkumar et al., 2010). Siderophore-producing endophytes have been reported from various plants, and the presence of these endophytes are known to enhance plant growth in low-nutrition environments (Idris et al., 2004; Barzanti et al., 2007; Sheng et al., 2008a, 2008b; Chen et al., 2010; Ma et al., 2011b; Zhang et al., 2011). This association of siderophore-producing endophytes with host plants can aid in ensuring growth and thriving nature of the plants in metal-polluted environments or in marginal lands.

Other than siderophore production, endophytic isolates solubilize mineral phosphate to aid plant growth under nutrient-limited environments. Phosphate-solubilizers are also relatively common, with 52% of endophytic bacteria of soybean plant known to produce this trait (Kuklinsky-Sobral et al., 2004). Interestingly, Puente et al. (2009) also found that most of the endophytic bacteria in cacti seeds have this trait, which assisted in the removal of key elements, such as P_2O_5, Fe_2O_2, K_2O, and MgO from rock substrates to support growth and development of cactus seedlings. Phosphate solubilization is achieved when endophytes produce substantial amounts of acid phosphatase, which is responsible in solubilizing phosphate. Solubilized phosphates are then ready and bioavailable for uptake by plants. *Piriformospora indica* is an example of phosphate-solubilizer that not only improves uptake of phosphate by the host plant, but also nitrate uptake as well (Oelmüller et al., 2009). In addition, some endophytes secrete low molecular mass organic acids that aid in heavy metal mobilization. This was demonstrated by endophytic *Gluconacetobacter diazotrophicus*, with the release of 5-ketogluconic acid, which dissolves ZnO, $ZnCO_3$, or $Zn_3(PO_4)_2$ as bioavailable forms for plant uptake (Saravanan et al., 2007). Other than the known mechanisms, actinobacterial endophytes were discovered to produce metal-mobilizing metabolites. These metabolites mobilize Zn and Cd and there are evidence or their mobility and gradual accumulation in leaves of *Salix caprea* (Kuffner et al., 2010). Phosphate-solubilizing endophytes, irrespective of their host plants, have the potential for metal-bioremediation due to their ability to mobilize metals.

The metal-bioremediation process may also be enhanced by the intentional inoculation/introduction of endophytes (usually from a phytoremediator plant) into new plants to remove toxic metals. This has been reported by several studies. Inoculation with the endophytic bacterium *Pseudomonas* sp. A3R3 increased Ni content in *Alyssum serpyllifolium* (Ma et al., 2011a). Chen et al. (2010) reported that inoculation with heavy-metal-resistant endophytic bacteria also enhanced the uptake and accumulation of Cd in leaf, stem, and root tissues of *Solanum nigrum* L.. Inoculations with Cd-resistant endophyte *Sanguibacter* sp. to *Nicotiana tabacum*, increased accumulation of Cd in shoot tissues (Mastretta et al., 2009). Similarly, Sheng et al. (2008b) reported an increase in Pb uptake in *Brassica napus* inoculated with Pb-resistant endophytic bacteria.

The endophytes can also be used independent of their host plant. Live or dead cells of the endophytes can be applied for metal removal via biosorption or bioaccumulation. For example, tropical endophyte *Lasiodiplodia* sp. MXSF31, from the hyperaccumulator *Portulaca oleracea*, is resistant to Cd, Pb, Zn (Deng et al., 2014). When applied as cell biomass for metal biosorption (independent of host plant), the isolate was able to absorb 3.0×10^3, 1.1×10^4, and 1.3×10^4 mg kg^{-1} of Cd, Pb, and Zn, respectively. The cell biomass allows metal cations to bind to the functional groups (e.g., amino, carbonyl, hydroxyl, sulfonate, benzene ring) present on the cell walls (Zahoor et al., 2017). As such, biosorption can occur as metal cations are bound to cell biomass and removed alongside the cell biomass (fungal biosorbent). Similar observations were reported by Sim et al. (2018, 2019), where endophytes from *Phragmites* demonstrated metal removal independent of the host plant.

In many of the studies, it was revealed that live cells could typically accumulate more toxic metal cations than the dead cells (Deng et al., 2014). Several mechanisms have been demonstrated; compartmentalization, bioaccumulation, and even methylation of metals by live cells. The drawback, however, is that live cells are susceptible to metal toxicity and environmental parameters, which may influence metal removal efficacy. For dead cells, the binding of metal cations to the surface of cell biomass is a passive process and is dependent on the saturation limits of the functional groups on the surface of the cells. With more binding occurring, more rapid metal uptake is observed, but there are also less available sites to further bind metal cations. To address saturation limits, attempts have been made to modify the density of these functional groups to enhance the biosorption efficiency for metal ions. As such, endophytic fungi are fashioned into desirable biosorbents (i.e., more functional groups as binding sites) via chemical modifications. Chemical or heat treatments on the biomass of endophytes have been found to enhance metal biosorption efficacy. These chemical or heat treatments are often applied prior to the use of biomass as biosorbents, thus they are often referred to as pretreatments. One example of successful pretreatment is the use of biomass from *Fusarium* sp. This endophyte, isolated from mangrove environments, was dried and powdered prior to pretreatment with various chemical agents (e.g., acetic acid, formaldehyde, methanol). As a result of the pretreatment, *Fusarium* cells have better affinity toward the uptake of toxic metals such as uranium in wastewater (Chen et al., 2014). Pretreated as well as non-treated cells can have improved metal uptake as well when other parameters are optimized (e.g., contact time, pH, initial metal ion concentrations) (Chen et al., 2014).

There are, however, conflicting views on whether metal removal or tolerance is more prominent in endophytes from environments with predisposing factors, such as polluted sites or localities with constant exposure to high metal residues. There are studies, which supported the findings that endophytes from polluted environments have inclination toward better efficiency in metal removal. For example, the mangrove ecosystem where metal accumulation occurs. The mangrove plant *Kandelia candel* was found to host the endophytic fungus *Purpureocillium* sp. A5 (Gong et al., 2017). The association of endophytic *Purpureocillium* with the host plant was beneficial as the isolate aided in reducing the uptake of toxic Cu by the host plant. This was achieved by having *Purpureocillium* sp. secrete compounds to enhance Cu-complexation in the soil (via carbonate-bound Cu, Mn–Fe complexes Cu, organic-bound Cu). As such, there is less bioavailable Cu for uptake by plants, thus reducing toxicity to plants. Another example is the association of Ni-resistant *Herbaspirillum seropedicae* with the host plant *Lolium perenne*, which resulted in a significant decrease of nickel concentrations in shoot and root tissues (Lodewyckx et al., 2001). Similarly, Ni and Cd, and Cu uptake by plants was significantly reduced with the presence of endophytic bacteria *Methylobacterium oryzae* and *Burkholderia* sp., and *Pantoea* sp. Jp3-3 in tomato and guinea grass, respectively (Madhaiyan et al., 2007; Huo et al., 2012).

On the contrary, evidence exist as well, that endophytes for metal removal or with metal-tolerant traits need not necessarily originate from hosts or hyperaccumulator plants from polluted environments. Instead, non-metal polluted host plants have also been found to harbor metal-resistant endophytes (Moore et al., 2006). These are demonstrated by bacterial endophytes (e.g., species of *Arthrobacter*, *Bacillus*, *Clostridium*, *Enterobacter*, *Microbacterium*, *Paenibacillus*, *Pseudomonas*, *Xanthomonas*, *Staphylococcus*, and *Stenotrophomonas*), as well as fungal endophytes (e.g., species of *Alternaria*, *Aspergillus*, *Microsphaeropsis*, *Mucor*, and *Phoma*) isolated from the various non-hyperaccumulator plants. This is further supported by findings that many metal-resistant endophytes have also been isolated from non-hyperaccumulating plants such as *Arabis hirsuta*, *Acacia decurrens*, and *Symplocos paniculata*.

2.6.2 Hydrocarbon Bioremediation

Hydrocarbon contamination in the tropics is a growing problem. The bioremediation of hydrocarbon is therefore important to remove toxic hydrocarbon compounds in a more environmentally friendly and sustainable approach. Endophytes have been discovered to have the potential to remove hydrocarbon by degrading the hydrocarbons. For endophytes, hydrocarbon removal is often performed in association with plants via phytoremediation. It has been discovered that the presence of endophytic oil-degrading bacteria in host plants, not only enhance the degradation of hydrocarbon (~78% degradation), but also improved plant growth. This was reported when endophytic *Acinetobacter* sp. strain BRSI56 and *Pseudomonas aeruginosa* strain BRRI54 were inoculated to the grasses *Brachiaria mutica* and *Leptochloa fusca*. The hydrocarbon degradation efficiency achieved by grasses with endophytes was significantly higher than degradation efficiency by grasses without strains BRSI56 and BRRI54 (Fatima et al., 2016). Similarly, Siciliano et al. (2001) also observed the role of endophytic microbial communities in performing phytoremediation of petroleum-contaminated soils.

Endophytic species from non-hydrocarbon polluted sites have also been found to harbor hydrocarbon-degrading endophytes. They have been isolated from *Cerrado* plants found growing in nutrient-poor soils, as well as common plants like *Hibiscus rosasinensis*. These endophytes demonstrate capacity to degrade the different fractions of hydrocarbon (e.g., diesel oil, gasoline, petroleum) and polythene suggesting a potential use for hydrocarbon bioremediation (Germaine et al., 2009). To enhance the bioremediation potential of endophytic bacteria, genetic engineering has been employed. For example, *Burkholderia cepacia* L.S.2.4 is a genetically-engineered bacterium carrying a pTOM toluene-degradation plasmid. This pTOM plasmid was derived from *B. cepacia* G4, an endophyte from the yellow lupine that is used in phytoremediation of toluene (Barac et al., 2004). The degradation of toluene was significantly higher when the recombinant strain was used (up to 50%–70%).

Similarly, inoculation of pea (*Pisum sativum*) with genetically-engineered *Pseudomonas putida* VM1441(pNAH7) degraded 2,4-dichlorophenoxyacetic acid effectively (up to 40% degradation) (Germaine et al., 2009).

2.6.3 Other Xenobiotic Pollutants

There are other xenobiotic pollutants in the environment that can be degraded or removed by tropical endophytic species. Endophytes have been explored for the removal of dyes, polyester polyurethane, and phenanthrene. Application of endophytic fungi *Phlebia* sp. and *Paecilomyces formosus* successfully degrade Reactive Blue 19 and Reactive Black 5 textile dyes. This was attributed to the extracellular laccase produced, responsible for dye decolorization (Bulla et al., 2017). Ting et al. (2016) also found endophytic *Diaporthe* sp. from *Portulaca* sp., a common weed in Malaysia, which has degradation activities towards triphenylmethane dyes. For polyester polyurethane (PUR), the main component of plastic, several endophytic fungi have demonstrated the ability to degrade PUR in both solid and liquid suspensions. The endophytes with PUR-degradation are *Pestalotiopsis microspora*, which utilizes PUR as the sole carbon source due to degradation by serine protease. For phenanthrene, positive degradation activities were demonstrated by *Ceratobasidum stevensii* from host plant *Bischofia polycarpa* (Dai et al., 2010).

There is also a growing interest in the potential role of endophytes in reducing air pollution. While there are limited studies on this, hypothesis arises that endophytes may have roles in curbing air pollution. This is possible either by having pollutants in the air absorbed into plants and subsequently detoxified (i.e., via sequestration, transformation, degradation); or by the action of the endophytic microbiome in producing volatile compounds to inhibit airborne microbes and purify the air (Wolverton, 2008; Berg et al., 2014). More studies are therefore required to determine the role of endophytes in managing air pollution.

Valuable Endophytic Species from the Tropics

3

3.1 ENDOPHYTIC ACTINOBACTERIA

Endophytic actinobacteria exist in tissues of healthy plants. They are ubiquitous as they can be found in almost any tropical plant; from forest native plants to cultivated crop plants. Endophytic actinobacteria grew important and attracted significant interest in recent years, primarily due to their production of secondary metabolites with various biological activities. Metabolites from actinobacteria have demonstrated the following properties; antibiotic, anti-tumor, anti-infection, enzyme-producer, and plant growth promoter (Strobel et al., 2004; Qin et al., 2009, 2012). The most predominant species are from the genera *Streptomyces*, followed by *Micromonospora*, *Microbispora*, *Nocardia*, *Nocardioides*, and *Streptosporangium*. Most of these endophytic actinobacteria enhance the growth of host plants and improve tolerance to stress such as biotic stress (Naik et al., 2008). Some rare genera (i.e., *Sphaerisporangium* and *Planotetraspora*) have also been reported from tropical plants and are reported as endophytic actinobacteria for the first time. Other rare genera that are becoming increasingly prevalent in tropical forests include *Actinocorallia*, *Blastococcus*, *Dietzia*, *Dactylosporangium*, *Jiangella*, *Oerskovia*, *Promicromonospora*, and *Saccharopolyspora*. Similar to bacterial and fungal endophytes, endophytic actinobacteria have also shown characteristics as latent pathogens. Species of *Frankia* and *Streptomyces scabies* are two examples as they could interact with their host plant to either express pathogenic or endophytic associations (Benson and Silvester, 1993; Doumbou et al., 1998).

3.2 *DIAPORTHE AND PHOMOPSIS* SPECIES

The genus *Phomopsis* has approximately 1000 species of ascomycetes. *Phomopsis* is the anamorphic phase while the teleomorphic phase is the genus *Diaporthe* (Pandey et al., 2003; Murali et al., 2006). The *Diaporthe-Phomopsis* species are relatively dominant in many endophytic communities and is also ubiquitously found in almost any host plants in the various geographical regions, including the mangrove swamps in the tropics (Rhoden et al., 2012; Sebastianes et al., 2013; Ribeiro et al., 2018). They are also known to exist not only as non-pathogenic endophytes, but also pathogenic forms as well as saprobes (Gomes et al., 2013; Dissanayake et al., 2015). *Diaporthe-Phomopsis* species are highly valued as they produce metabolites with desirable bioactivities such as antibacterial, antifungal, antiviral, antimalarial, anticancer (cytotoxic) and herbicidal properties (Gomes et al., 2013). These metabolites are primarily polyketides, terpenoids, peptides, and quinone derivatives.

For *Diaporthe* sp., one of the most potent polyketides is the cytoskyrins. Cytoskyrins are effective antibacterial agents as they are DNA-damaging agents, most effective toward gram-positive bacteria. They also show strong cytotoxic activities (Bady et al., 2000; Agusta et al., 2006). In addition, *Diaporthe* species also produce aromatic polyketides (e.g., diaporine, diaporthemin, and xanthones) and anthraquinone and bixanthanoe derivatives. These were determined from chemical investigations using endophytic *Diaporthe melonis* from *Annona squamosa* (Ola et al., 2014). Dihydroanthracenone atropodiastereomers, diaporthemin A, and the dihydroanthracenone flavomannin-6,6'-dimethylether are antibacterial, especially flavomannin-6,6'-dimethylether, against *Streptococcus pneumoniae* and the methicillin-resistant *Staphylococcus aureus* (Ola et al., 2014). *Diaporthe* sp. also produces natural xanthones, which are antimalarial. This was observed in *Diaporthe* strain CY-5286 from mangrove plants, which produces the valuable xanthone described as Dicerandrol D (Calcul et al., 2013).

Phomopsis species are equally valuable as sources of beneficial bioactive metabolites. The profile of metabolites produced by *Phomopsis* is similar to metabolites produced by *Diaporthe*, which includes a variety of metabolites such as dicerandrols, terpenoids, peptides, and quinones. The dicerandrols produced are types A, B, and C (by *Phomopsis longicolla* and *Phomopsis* sp. HNY29-2B), which are highly antibiotic and cytotoxic (Wagenaar and Clardy, 2001; Lim et al., 2010; Ding et al., 2013). The other compound, phomoxanthones, is also linked to dicerandrols. Phomoxanthones A and B, and other similar phomaxanthone compounds such as deacetylphomoxanthone A and B, and penexanthone A or monodeacetylphomoxanthone B, have also been found to be highly similar to the dicerandrols produced by *Diaporthe* sp. These phomoxanthones, especially phomoxanthones A and B, are highly desirable as they have demonstrated antifungal, antimalarial, antitubercular, and cytotoxic activities (Isaka et al., 2001; Rukachaisirikul et al., 2008; Lim et al., 2010; Choi et al., 2013a, 2013b; Rönsberg et al., 2013). *Phomopsis* sp. also produce terpenoids just as *Diaporthe* species. Examples of some of the terpenoids produced are norsesquiterpene γ-lactones and oblongolides A–M. Terpenoids have low toxicity profile but have been effectively used in cancer treatment (Anupam et al., 2011). Terpenoids also have antibacterial, antifungal, and herbicidal activities (Dai et al., 2005; Dettrakul et al., 2003). Some of the terpenoids produced include norsesquiterpene γ-lactones, oblongolides A–M. Other than terpenoids, peptides are also common compounds produced by *Diaporthe*. Examples of peptides produced are cyclosporin, fingolimod, and emodepside. Peptide production in *Diaporthe* is, unfortunately, low in numbers. They find applications in agricultural and pharmaceutical industries, primarily due to their antimicrobial nature. Exumolide A is applied for antimicroalgal and cytotoxic treatments (Yang et al., 2011), and Fusaristatin A for antitumor treatments (Lim et al., 2010). The last group is the quinones. Quinones are cyclic polyketides, which are typically pigmented. They have been reported to have antibiotic, cytotoxic, phytotoxic, and anti-HIV properties (Evidente et al., 2011; Yang et al., 2013). Examples of well-known quinones are fusarubin, altersolanols A and B, and phomonaphthalenone A (Evidente et al., 2011; Yang et al., 2013).

3.3 *LASIODIPLODIA* SPECIES

The endophytic *Lasiodiplodia* sp. is researched primarily for its volatile compounds. An extensive study was conducted by Qian et al. (2014), who isolated *Lasiodiplodia* endophytes from *Viscum coloratum* (mistletoe). *V. coloratum* is an important medicinal herb used for many years in traditional Chinese medicine (Qian et al., 2014). It was discovered that the flowers of *V. coloratum* host the endophyte *Lasiodiplodia* sp. ME4-2, which produces major volatile compounds. Among the many compounds, four volatile compounds with relatively high concentrations were identified, which are indole-3-carbaldehyde, indole-3-carboxylic acid (ICA), cyclo-(Trp-Ala), 2-phenyletha-nol, and mellein. The functions of these compounds are gradually established, with ICA proposed as accelerating callose accumulation in response to infec-tion by pathogens (biotic stress). Mellein also has a role in promoting plant growth. For 2-phenylethanol, this is a volatile compound with similar profiles to essential oils in flowers.

3.4 *MUSCODOR* SPECIES

Muscodor species are common endophytes found from tropical trees and vines in the countries in the tropics. Studies have generally reported on metabo-lites of *M. roseus*, *M. vitigenus*, and *M. crispans*, from countries in Thailand, Australia, and in Central America. Endophytic *Muscodor* species are isolated from *Ananas ananassoides* (Mitchell, 2008), *Erythophelum chlorostachys* and *Grevillea pteridifolia* (Worapong et al., 2002), *Cinnamomum zeylani-cum* and *Myristica fragrans* (Worapong et al., 2001; Sopalun et al., 2003), and *Paullinia paullinioides* (Daisy et al., 2002). The volatile organic compounds (acids, alcohols, esters, ketones, 3-methyl butyl acetate or isoamyl acetate, and lipids) produced by *Muscodor* species may be a mixture of toxic volatile organic compounds that inhibits plant and human pathogens (Ezra et al., 2004; Strobel, 2006; MacíasRubalcava et al., 2010; Macías-Rubalcava and Sánchez-Fernández, 2017a, 2017b).

3.5 *TRICHODERMA* SPECIES

Trichoderma species is a common mycoparasite found inhabiting many differ-ent host plants. They can be isolated from almost any plant part such as roots, stems, fruits, and tree trunks (Bailey et al., 2006, 2008). Some of the endo-phytic *Trichoderma* species discovered are *T. evansii* (Samuels and Ismaiel, 2009), *T. koningiopsis* (Samuels et al., 2006a), *T. martiale* (Hanada et al., 2008), *T. ovalisporum* (Holmes et al., 2004), *T. paucisporum* (Samuels et al., 2006b), *T. stromaticum* and *T. theobromicola* (Samuels et al., 2000). In Malaysia, in the natural forest and plantation sites in Sabah and Sarawak, the diversity of *Trichoderma* species was profiled. Most of the isolates were discovered to be *Trichoderma asperellum*, *T. asperelloides*, *T. afroharzianum*, *T. guizhouense*,

T. strigosum, T. reesei and *T. virens*. Among these, *T. asperellum/T. asperelloides, T. harzianum* and *T. virens*, are the most prevalent (Cummings et al., 2016). *T. asperellum* and *T. harzianum* were also the most prevalent species found in tropical parts of China (Sun et al., 2012).

Most of the tropical *Trichoderma* endophytes are isolated from crops. They have been isolated from roots of *Coffea arabica* in Ethiopia (Hoyos-Carvajal and Bissett, 2011; Samuels et al., 2012; Mulaw et al., 2013), from cocoa (Mpika et al., 2009), and from banana (Xia et al., 2011). Endophytic *Trichoderma virens* and *T. harzianum* were dominant in the cocoa rhizosphere (from Ivory Coast) (Mpika et al., 2009), while *T. asperellum, T. virens,* and *T. harzianum* (as *Hypocrea lixii*) were recovered from banana roots in China (Xia et al., 2011). *Trichoderma* are naturally fast growing and are extensive and efficient colonizers. Their colonization within tissues is, however, limited to the first few layers of the root cortex, as plant defense responses restrict their internal colonization and growth (Harman et al., 2004; Yedidia et al., 1999). The association of *Trichoderma* with host plants has rendered benefits to the host plant. One of the most pronounced benefits is on root growth, nutrient uptake, and general plant development (Harman, 2011). *Trichoderma* also protect host plants from pathogens as *Trichoderma* are efficient mycoparasite. They could, therefore, inhibit the growth of pathogens, either through mycoparasitism (Druzhinina et al., 2011), production of inhibitory compounds (Reino et al., 2008), competitive exclusion for nutrients and space (Elad, 1996; Howell, 2003) or through induced host plant defenses (Harman et al., 2004; Hermosa et al., 2012). As such, *Trichoderma* has become one of the more important bioagents for development into biocontrol and biofertilizer agents for the control of pathogens (Harman, 2000; Howell, 2003; Benítez et al., 2004). For example, endophytic *Trichoderma* isolates from *Coffea arabica* showed potential for the control of major plant pathogens (Hoyos-Carvajal and Bissett, 2011; Samuels et al., 2012; Mulaw et al., 2013).

3.6 *XYLARIA* SPECIES

Xylaria species are also known to exist as endophytes, though at times as saprophytes. To date, there are more than 100 species of *Xylaria* (Webster and Weber, 2007). Species of *Xylaria* is relatively dominant in endophytic communities as their recovery rate is frequent and high among other endophytic species. They are found in almost any tropical plants (Arnold and Lutzoni, 2007; Hyde and Soytong, 2008) as their generalist nature enables efficient colonization in various host plants. *Xylaria* sp. are not just inhabitants of vascular plants (e.g., monocots, dicots, conifers, ferns, and lycopsids), but also non-vascular plants (e.g., liverworts) (Davis et al., 2003). The bioactive metabolites produced by *Xylaria* sp. include alkaloids, cytohalasins, terpenoids (e.g., eudesmanes, guaianes), mellein, polyketides, and volatile organic compounds (e.g., alcohols, esters, and sesquiterpenoids) (Song et al., 2014; Sánchez-Ortiz et al., 2016). Due to their bioactivities, endophytic *Xylaria* sp. finds applications as insecticides, fungicides, and herbicides, among others. Their applications in the agriculture sector are primarily as biocontrol agents and for biopesticide use, and their use in medicine is for the treatment of infectious and non-infectious diseases (as antibacterial, antifungal, and antimalarial agents).

Commercialization of Endophytes from the Tropics

4

4.1 INTRODUCTION

To date, endophytes from the tropics have been researched and developed as commercial products. The commercialized products can be in the form of whole cells, or their secondary metabolites or derivatives. Whole endophytic cells can be commercialized as bioagents for various applications, typically for field application to promote plant growth, and to enhance tolerance of host plants toward abiotic and biotic stress (suppressing disease development). The common bioformulation developed is liquid formulation. Liquid formulations are usually applied onto the seeds (via spraying), by commercial seed treatment companies or by farmers themselves. The advantage of using endophytes is that endophytes will remain dormant on the seed, and upon seed germination, endophytes will establish an association with the germinated seedlings. Endophytes are therefore useful to improve the response of agricultural crops in relation to climate changes (i.e., water, drought, and salination stresses). In addition to liquid formulations, solid formulations are also useful, with powder, pellet, encapsulated beads, and dust forms attempted and investigated (Figure 5).

Examples of commercialized bioagent for field applications is the endophytic *Piriformospora indica*. *P. indica* was developed as powder formulations under the trade name "ROOTONIC" and is used in field trials in India to promote root growth (Shrivastava and Varma, 2014). In addition, there are also endophytes that have been developed to confer tolerance to abiotic stress. These are typically aimed at stress factors such as salt and heat tolerance (Rodriguez et al., 2008; Lugtenberg et al., 2016). Endophytes developed for tolerance

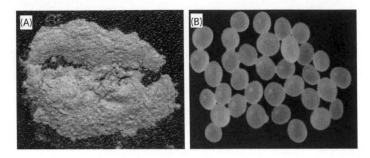

FIGURE 5   Examples of bioformulations of endophytic isolates in (A) powder form and (B) encapsulated alginate beads.

adaptations that have been commercialized includes *BioEnsure* R-*Corn* and *BioEnsure* R-*Rice*. *BioEnsure* R-*Corn* is reportedly able to increase corn yield to 25%–80% under stressful heavy drought conditions. While on mild or low drought stress conditions, the yield was improved by 7%. The application of *BioEnsure* R-*Corn* allows the plant to tolerate drought conditions as water usage is 25%–50% lesser under normal conditions as well as under low drought stress. Similarly, *BioEnsure* R-*Rice* promoted yield as well, with a 25%–40% decrease in water use.

The commercialization of secondary metabolites is less developed compared to whole cells. Nevertheless, there are several commercialized secondary products from endophytes. They include pigments and volatile organic compounds, which are used in the food industries. Several pigments produced by endophytes have been commercialized. *Monascus* spp. produces many pigments. Some other pigments commercialized are lycopene and β-carotene from *Blakeslea trispora*, riboflavin from *Ashbya gossypii*, and Arpink red™ from *Penicillium oxalicum* (Dufosse et al., 2014).

Applications of volatiles have also been explored. Volatile compounds from *Muscodor albus* have been used for the treatment of soil fumigation, seed treatments, and on post-harvest commodities such as fruits and cut flowers (Ezra et al., 2004). Other than *Muscodor*, endophytic *Daldinia concentrica* also produces a mixture of volatile organic compounds that are applicable to control post-harvest diseases in dried fruits and other plant products. In principle, the development of commercialization for pigments and volatiles from endophytes meant for food, coloring, flavoring, and preservative purposes, have to meet the requirements on the FDA GRAS list.

Challenges in Endophytic Research in the Tropics

5

5.1 CHALLENGES AND LIMITATIONS

The research on endophytes in the tropics is not without challenges. There are fundamental complexities to the approach, including analysis and actual representation of the endophyte communities, to the challenges in identifying valuable compounds and upscaling their production for applications. One of the most fundamental challenges in endophytic research is on sampling. For sampling, the challenge lies with gathering sufficient number of samples (sample size) to be statistically significant as evidence to host- or habitat-preference analysis (Arnold, 2007). Sample size that is insufficient may lead to non-statistical observations, offering a conclusion that the endophytes are "host-generalist" when they could have been otherwise (Higgins et al., 2014). The other challenge in sampling is that most of the sampling is a "one-time" sampling procedure, which at best is possibly a snap-shot record of the endophyte community. This approach lacks information on the interactive dynamics of the endophytic communities (Suryanarayanan, 2013), which impacts the species richness and abundance, and subsequently the endophyte diversity in a specific habitat. Periodic sampling may mitigate this, but the frequent sampling may impose a costing burden. As such, all aspects of sampling, i.e., the number of samples, geographical location, selection of host plants, and sampling method are often inadequate to gather a wholistic representation of the entire endophyte community. Sampling is also important as it determines the types of endophytes and their bioactivities for bioprospecting. And these factors, if not planned strategically, may offer a bias outlook as they either overestimate or underestimate the endophyte communities. As such, the challenge is in adopting an approach to assists in actualizing the results to represent the natural distribution of endophyte communities as accurately as possible.

In addition to sampling, the diversity estimates are also a challenge, given that there are various methods for sequencing and taxa assignment. Typically, the analysis of species richness and diversity excludes singletons (OTUSs occurring once), so the conclusions are often relatively inconclusive as fewer than half of the OTUs are accounted (Arnold et al., 2000; Davis and Shaw, 2008; Gazis and Chaverri, 2010). In recent years, culture-free methods via PCR or cloning (Arnold et al., 2007; Higgins et al., 2011) and next-generation sequencing (Jumpponen and Jones, 2009) have been used to identify phylogenetically linked endophytes to their common genotypes or phylotypes (Arnold et al., 2007; Higgins et al., 2011; Gazis et al., 2012). The fact that some endophytes exist as complex anamorph and teleomorph nature further complicates diversity estimates. Therefore, it is important that the life cycle of the endophytes is determined. This is also to establish their anamorph-teleomorph associations, and their survival and dispersal outside of the host plants (Bills et al., 2012).

The estimation of endophytic occurrence and incidence in various host plants is often performed based on the presence/absence data of endophytic species (occurrence) or based on the abundance of endophytic fungi in each sample (incidence) (Arnold et al., 2000, 2001; U'Ren et al., 2010). The estimation based on occurrence or incidence, will ultimately yield a different overview on the composition of endophytic communities. For example, the estimation using relative abundance (incidence) rather than presence/absence of endophytes (occurrence), would lead to conclusions that tropical fungal communities may demonstrate host affinity (Higgins et al., 2014).

In addition to the challenges related to endophyte fundamental research, there are also challenges in discovering and applying endophytes or their compounds/metabolites for use. For example, endophytic biocontrol agents have to be maintained at certain inoculum level when applied to the field. This is to provide sufficient inoculum level to render inhibitory effects on pathogens. It has been discovered that the inoculum level typically diminishes over time. This was observed for *Trichoderma* sp., where spore suspension (10^7 spores ml^{-1}) applied to vegetable crops allowed recovery of ~90% of inoculum 7 days after treatment. This gradually decreased to 30%–40% recovery by 28 days after spraying (Gange et al., 2012). When developing endophytes as biocontrol agents, these endophytes may elicit different responses in different host plants. Not all host plants respond positively to endophyte infection. In some cases, endophytes as biocontrol agents were reported to implicate plant growth under certain circumstances as they have also been shown to be latent pathogens. When hosts are exposed to certain circumstances or environmental conditions that are highly stressful, pathogenesis related to fungal stress or mitogen-activated protein kinases (e.g., sakA) may be triggered (Eaton et al., 2011).

There are also challenges in bioprospecting for compounds/metabolites produced by endophytes. While the tropics especially tropical forests are favored as areas of high biodiversity ("biodiversity hotspots"), with possible novel compounds discovered, the diversity can be enormous (Strobel, 2006). As such, the diversity of the culture library is extremely large (Strobel and Daisy, 2003; Arnold and Lutzoni, 2007), which makes the task of selecting, identifying, and bioprospecting the compounds/metabolites a challenge. Once the compounds have been identified and characterized, the next task is to develop their bioprocess production and scale-up. These have met with little success as there are numerous parameters to be considered with design and optimization processes to be configured. Some of the parameters for consideration include; cultivation times, temperature, solubility of oxygen, temperature variation, viscosity, and pH. Similarly, the upscaling for volatile organic compounds are influenced by inefficient energy supply during cellulose degradation in fungi, may impact growth and production of metabolites/compounds (Stadler and Schulz, 2009).

5.2 REMEDIAL STRATEGIES TO CHALLENGES

The challenges in endophytic research are real and present and demand remedial strategies. For endophyte studies in the tropics, the estimation of diversity, occurrence, and abundance is at best, an educated approximate based on existing resources and tools for estimation. With the integration of genomics and sequencing technologies, the diversity estimates are expected to be more inclusive, considering non-culturable endophytes are interpreted as well. The tropical regions are consistently humid and wet, and endophyte infection is expected to remain high. But this may result in an overgeneralization of the abundance of endophytes in a particular environment. Therefore, it is best that diversity studies are coupled with effective ecological mapping and molecular characterization, to provide a true representation of the endophyte richness in the tropics.

For improvements to applications of endophytes or their metabolites, yield and productivity can be improved via co-culture fermentation, process optimization, and genetic engineering. Co-culture fermentation involves the cultivation of more than a single species/isolate within the same culture vessel or culture conditions (Bertrand et al., 2014). This approach adopts the concept that the interaction or coexistence of two different strains could trigger the expression of secondary metabolite gene clusters differently, resulting in the production of secondary metabolites with high bioactivity. It is theorized that these gene clusters may be activated as a response to competition or the mere assurance of natural coexistence with another strain. This was observed by Bhalkar et al. (2016) when endophytic *Colletotrichum fructicola* SUK1 and *Corynespora cassiicola* SUK2 (isolated from plant *Nothapodytes nimmoniana* (Grah.) Mabb. (Ghanera)), were establish as co-cultures and they synthesized CPT at >1.4 fold higher than amount derived from single-cultures. Co-culture fermentation has also been beneficial in stimulating the production of hydrocarbon by endophytic *Gliocladium roseum* upon co-culturing with *Escherichia coli*. The production of hydrocarbon was enhanced 100-fold higher than when single *G. roseum* culture was used (Ahamed and Ahring, 2011).

Co-culturing can also be done with more than two isolates. The co-culturing of *Paraconiothyrium* SSM001 (isolated from Taxus (yew) trees) with *Alternaria* sp. and *Phomopsis*, that were also isolated from the yew plant, resulted in an eight-fold increase in taxol yield. This is much higher than when only two isolates (SSM001 and *Alternaria*) were used, in which the taxol production increased by three-fold (Soliman and Raizada, 2013). The results from this study also suggested that co-culturing with endophytes from the same host plant might be beneficial, as they tend to stimulate the biosynthesis of secondary metabolites. This approach is worth the exploration to enhance the production of antibiotics as presence of another other isolate is expected to induce the production of the compounds.

Manipulating several other factors can also assist in optimizing secondary metabolite production. This optimization process manipulates culture media composition (ratio of nutrients, pH) and incubation conditions (temperature, aeration, period of incubation) (Elmoslamy et al., 2017). This has been proven successful for several endophytes. For example, the production of diuron-degrading enzymes were enhanced when the endophyte *Neurospora intermedia* DP8-1 (from sugarcane roots) was fermented under optimum carbon source, pH, temperature, and inoculum and initial diuron concentrations. This resulted in 99% of diuron degradation (Wang et al., 2017). Similarly, production of polyketides by *Fusarium solani* (from *Ferocactus latispinus*, Devil's Tongue Barrel) was also enhanced when carbon/nitrogen ratio and pH conditions were optimized (Gracida-Rodríguez et al., 2017). In recent years, optimization also includes incubation in dark and light conditions. As endophytes, dark conditions are typically favored compared to light treatments. This may be attributed to the nature of endophytes existing in tissues that are usually shielded from light (under dark conditions). This was observed when *Paraconiothyrium* SSM001 was incubated in dark conditions where the production of taxol is higher than cultures incubated in light. Light exposure is postulated to have a repressive effect on the genes involved in taxol biosynthesis (Soliman and Raizada, 2018).

There are other strategies adopted too, which include genetic engineering. This is usually for selected important endophytes, as genetic engineering will allow genetic manipulation for specific expressions on the production of desirable compounds. The early endophytic strains selected for genetic engineering were taxol-producing endophytic fungi. To enhance taxol production, genetic engineering via random mutagenesis, genome shuffling, and gene overexpression were performed (Ahamed and Ahring, 2011; El-Gendy et al., 2016). Protoplast fusion increased taxol yield by 20%–25% compared to the parental strains (Zhao et al., 2008). Genome shuffling is also effectively used to engineer *Phomopsis* sp. (endophyte of mangrove plants) to produce deacetylmycoepoxydiene (DAM), an antitumor agent. The engineered mutants produce >200-fold increase in yield after two cycles of genome shuffling (Wang et al., 2016). A similar taxol-producing endophyte *Ozonium* sp. EFY-21 (isolated from *Taxus chinensis* var. *mairei*, Chinese yew), was engineered to overexpress the enzyme gene, TS. The enzyme TS performs the first catalysis of the taxol biosynthesis reaction, resulting in approximately five-fold increase in taxol production compared to the parent strain (Wei et al., 2012).

The remedial strategies proposed are not exhaustive, but are indications of what has been attempted and what could further be introduced as innovations to advance research on tropical endophytes. Selection of approaches is dependent on objectives of study.

Conclusions

6

The tropical regions clearly host a diverse plethora of endophytes, and these endophytes are rich in various valuable properties. The discovery of endophytes, their metabolites and amenability to improvements, present them as attractive alternative bioresources. The use of endophytes in many applications is essentially a more sustainable and environmental-friendly approach. The endophytes from the tropics have learned to adapt to the many stress factors in the tropics. Tropical conditions are peppered with abiotic stress (drought, water-logged, aerobic, anaerobic, salinity, pH stress, anthropogenic factors) and biotic stress (disease outbreak due to high humidity), thus the endophytes developed adaptations to these stress factors, resulting in the remarkable range of metabolites produced. The production of metabolites by endophytes is known to be influenced by chemical ecology attributed to the long-term interactions/coevolution of endophytes with its hosts and environment.

Endophytes from the tropics are therefore useful in almost any application. They are highly sought for agriculture, environmental bioremediation, biomedicine, bioenergy, and for biocatalysis. There is room for further development, especially in perfecting the development of these endophytes or their derivatives for applications. The aspects for improvement include optimizing the biosynthesis process, scale-up development, and eradication of possible toxic metabolites coproduced with the desired metabolites so that greater applications can occur. Future studies can, therefore, be aimed toward uncovering biosynthesis mechanisms for valuable secondary metabolites and better approaches in discovering novel natural products.

It is also expected that new technologies such as DNA cloning (Guo et al., 2001; Seena et al., 2008), DGGE (Duong et al., 2006; Tao et al., 2008), T-RLFP (Nikolcheva and Barlocher, 2004, 2005), and shot-gun metagenomics (Kimura, 2006), could provide more information on unculturable endophytes, their metabolite-producing potential, and the endophyte microbiome in the endophyte-plant ecosystem. This would provide information on the endophytes, the host plants, and the different species growing in the different habitats. The production of secondary metabolites especially novel compounds (or lead compounds), can be further established via the metabolomics approach (Jewett et al., 2006).

To conclude, tropical plants are excellent reservoirs that host endophytes, many of which are undiscovered. These endophytes are potential sources of novel compounds. It is therefore beneficial to include endophytes into natural products discovery programs. A global initiative involving fungal taxonomists, ecologists, and natural product chemists is also expected to benefit endophyte research.

REFERENCES

Agusta, A., Maehara, S., Ohashi, K., Simanjuntak, P., Shibuya, H. 2005. Stereoselective oxidation at C-4 of flavans by the endophytic fungus *Diaporthe* sp. isolated from a tea plant. *Chemical and Pharmaceutical Bulletin* 53:1565–1569. doi:10.1248/cpb.53.1565

Agusta, A., Ohashi, K., Shibuya, H. 2006. Composition of the endophytic filamentous fungi isolated from the tea plant *Camellia sinensis*. *Journal of Natural Medicines* 60(3):268–272.

Ahamed, A., Ahring, B.K. 2011. Production of hydrocarbon compounds by endophytic fungi *Gliocladium* species grown on cellulose. *Bioresoure Technology* 102(20):9718–9722. doi:10.1016/j.biortech.2011.07.073

Almeida, A.C., Ortega, H., Higginbotham, S., Spadafora, C., Arnold, A.E., Coley, P.D., Kursar, T.A., Gerwick, W.H., Cubilla-Rios, L. 2014. Chemical and bioactive natural products from *Microthyriaceae* sp., an endophytic fungus from a tropical Grass. *Letters in Applied Microbiology* 59:58–64. doi:10.1111/lam.12245

Aly, A.H., Debbab, A., Proksch, P. 2011. Fungal endophytes: Unique plant inhabitants with great promises. *Applied Microbiology Biotechnology* 90:1829–1845.

Anupam, B., Shamima, A., Nikoleta, B., Marjorie, P. 2011. Triterpenoids as potential agents for the chemoprevention and therapy of breast cancer. *Frontiers in Bioscience*, 16:980–996.

Araujo, J.M., Silva, A.C., Azevedo, J.L. 1999. Isolation of endophytic actinomycetes from roots and leaves of maize (Zea mays L.). *Brazilian Archives of Biology and Technology* 43(4). doi:10.1590/S1516-89132000000400016

Arnold, A.E. 2007. Understanding the diversity of foliar endophytic fungi: Progress, challenges, and frontiers. *Fungal Biology Reviews* 21:51–66.

Arnold, A.E. 2008. Hidden within our botanical richness, a treasure trove of fungal endophytes. *Plant Press* 32: 13–15.

Arnold, A.E., Henk, D.A., Eells, R.A., Lutzoni, F., Vilgalys, R. 2007. Diversity and phylogenetic affinities of foliar fungal endophytes in loblolly pine inferred by culturing and environmental PCR. *Mycologia* 99:185–206.

Arnold, A.E., Herre, E.A. 2003. Canopy cover and leaf age affect colonization by tropical fungal endophytes: Ecological pattern and process in *Theobroma cacao* (Malvaceae). *Mycologia* 95:388–398.

Arnold, A.E., Lutzoni, F. 2007. Diversity and host range of foliar fungal endophytes: Are tropical leaves biodiversity hotspots? *Ecology* 88:541–549.

Arnold, A.E., Maynard, Z., Gilbert, G.S. 2001. Fungal endophytes in dicotyledonous neotropical trees: Patterns of abundance and diversity. *Mycological Research* 105:1502–1507.

Arnold, A.E., Maynard, Z., Gilbert, G.S., Coley, P.D., Kursar, T.A. 2000. Are tropical fungal endophytes hyperdiverse? *Ecology Letters* 3:267–274.

Arnold, A.E., Mejıa, L.C., Kyllo, D., Rojas, E.I., Maynard, Z., Robbins, N., Herre, E.A. 2003. Fungal endophytes limit pathogen damage in a tropical tree. *Proceedings of the National Academy of Sciences USA* 100:15649–15654.

Arnold, A.E., Miadlikowska, J., Higgins, K.L., Sarvate, S.D., Gugger, P., Way, A., Hofstetter, V., Kauff, F., Lutzoni, F. 2009. Phylogenetic estimation of trophic transition networks for ascomycetous fungi: Are lichens cradles of symbiotrophic fungal diversification? *Systematic Biology* 182:314–330.

Assis, S.M., Silveira, E.B., Mariano, R., Menezes, D. 1998. Endophytic bacteria-method for isolation and antagonistic potential against cabbage black rot. *Summa Phytopathologica* 24(3/4):216–220.

Ateba Joël, E.T.A., Toghueo, R.M.K., Awantu, A.F., Mba'ning, B.M., Gohlke, S., Sahal, D., Rodrigues-Filho, E., Tsamo, E., Boyom, F.F., Sewald, N., Lenta, B.N. 2018. Antiplasmodial properties and cytotoxicity of endophytic fungi from *Symphonia globulifera* (Clusiaceae). *Journal of Fungi* 4:70. doi:10.3390/jof4020070

Azevedo, J.L., Maccheroni, Jr. W., Pereira, J.O., Araujo, W.L. 2000. Endophytic microorganisms: A review on insect control and recent advances on tropical plants. *Journal of Biotechnology* 3(1):40–65.

Azevedo-Silva, F., deCamargo, B., Pombo-de-Oliveira, M.S. 2010. Implications of infectious diseases and the adrenal hypothesis for the etiology of childhood acute lymphoblastic leukemia. *Brazillian Journal of Medical and Biological Research*, 43(3):226–229. https://doi.org/10.1590/S0100-879X2010007500011

Bacon, C., White, J. 2000. *Microbial Endophytes*. New York: Marcel Dekker Inc.

Bae, H., Sicher, R.C., Kim, M.S. et al. 2009. The beneficial endophyte *Trichoderma hamatum* isolate DIS 219b promotes growth and delays the onset of the drought response in *Theobroma cacao*. *Journal of Experimental Botany* 60:3279–3295.

Bailey, B.A., Bae, H., Strem, M.D., Roberts, D.P., Thomas, S.E., Crozier, J., Samuels, G.J., Choi, I.Y., Holmes, K.A. 2006. Fungal and plant gene expression during the colonization of cacao seedlings by endophytic isolates of four *Trichoderma species*. *Planta*, 224:1449–1464. https://doi.org/10.1007/s00425-006-0314-0

Bailey, B.A., Holmes, K.A. 2008. Antibiosis, mycoparasitism, and colonization success for endophytic *Trichoderma* isolates with biological control potential in *Theobroma cacao*. *Biological Control*, 46(1):24-35. https://doi.org/10.1016/j.biocontrol.2008.01.003

Balakumaran, M.D., Ramachandran, R., Kalaichelvan, P.T. 2015. Exploitation of endophytic fungus, *Guignardia mangifera* for extracellular synthesis of silver nanopartilces and their in vitro biological activities. *Microbiological Research*, 178:9–17. https://doi.org/10.1016/j.micres.2015.05.009

Ballio, A., Bossa, F., Di Giorgio, D. et al. 1994. Novel bioactive lipodepsipeptides from *Pseudomonas synringae*: The pseudomycins. *FEBS Letters* 355(1):96–100.

Baltruschat, H., Fodor, J., Harrach, B.D., Niemczyk, E., Barna, B., Gullner, G., Janeczko, A., Kogel, K.H., Schäfer, P., Schwarczinger, I. 2008. Salt tolerance of barley induced by the root endophyte *Piriformospora indica* is associated with a strong increase in antioxidants. *New Phytologist* 180(2):501–510. doi:10.1111/j.1469-8137.2008.02583.x

Bamford, P.C., Norris, G.L.F., Ward, G. 1961. Flavipin production by *Epicoccum* spp. *Transactions of the British Mycological Society* 44:354–356.

Banerjee, U.C., Scrivastava, J.P. 1993. Effect of pH and glucose concentration on the production of rifamycin oxidase by *Curvularia lunata* in batch reactor. *Journal of Biotechnology* 28:229–236.

Baraban, E.G., Morin, J.B., Phillips, G.M., Phillips, A.J., Strobel, S.A., Handelsman, J. 2013. Xyolide, a bioactive nonenolide from an Amazonian endophytic fungus, *Xylaria feejeensis*. *Tetrahedron Letters* 54:4058–4060.

Barac, T., Taghavi, S., Borremans, B., Provoost, A., et al. 2004. Engineered endophytic bacteria improve phytoremediation of water-soluble, volatile, organic pollutants. *Nature Biotechnology*, 22:583–588. https://doi.org/10.1038/nbt960

Barzanti, R., Ozino, F., Bazzicalupo, M., Gabbrielli, R., Galardi, F., Gonnelli, C., Mengoni, A. 2007. Isolation and characterization of endophytic bacteria from the nickel hyperaccumulator plant *Alyssum bertolonii*. *Microbial Ecology* 53:306–316.

Bashan, Y., Okon, Y. 1981. Inhibition of seed germination and development of tomato plants in soil infested with *Pseudomonas* tomato. *Annals of Applied Biology* 98:413–417.

Baute, M.A., Deffieux, G., Baute, R. et al. 1978. New antibiotics from the fungus *Epicoccum nigrum*. I. Fermentation, isolation and antibacterial properties. *Journal of Antibiotics* 31:1099–1101.

Bengyella, L., Iftikhar, S., Nawaz, K., Fonmboh, D.J., Yekwa, E.L., Jones, R.C., Njanu, Y.M.T., Roy, P. 2019. Biotechnological application of endophytic filamentous bipolaris and curvularia: A review on bioeconomy impact. *World Journal of Microbiology and Biotechnology* 35:69. doi:10.1007/s11274-019-2644-7

Benitez, T., Rincon, A.M., Limon, M.C., Codon, A.C. 2004. Biocontrol mechanisms of *Trichoderma* strains. *International Microbiology*, 7:249–260.

Benson, D.R., Silvester, W.B. 1993. Biology of *Frankia* strains, actinomycete symbionts of actinorhizal plants. *Microbiology Reviews* 57:293–319.

Berdy, J. 2005. Bioactive microbial metabolites. *Journal of Antibiotics* 58:1–26.

Berg, G., Hallmann, J., 2006. Control of plant pathogenic fungi with bacterial endophytes. In: Schulz, B., Boyle, C., Sieder, T. (eds). *Microbial Root Endophytes*. Springer, Berlin, Germany. pp. 53–69.

Berg, G., Mahmert, A., Moissl-Eichinger, C. 2014. Beneficial effects of plant-associated microbes on indoor microbiomes and human health? *Frontiers in Microbiology* 5:1–5.

Bertrand, S., Bohni, N., Schnee, S., Schum, O., Gindro, K. 2014. Metabolite induction via microorganism co-culture: A potential way to enhance chemical diversity for drug discovery. *Biotechnology Advances*, 32(6):1180–1204.

Bezerra, J.D.P., Santos, M.G.S., Barbosa, R.N., Svedese, V.M., Lima, D.M.M., Fernandes, M.J.S., Gomes, B.S., Paiva, L.M., Almeida-Cortez, J.S., Souza-Motta, C.M. 2013. Fungal endophytes from cactus *Cereus jamacaru* in Brazilian tropical dry forest: A first study. *Symbiosis* 60:53–63. doi:10.1007/s13199-013-0243-1

Bhalkar, B.N., Patil, S.M., Govindwar, S.P. 2016. Camptothecine production by mixed fermentation of two endophytic fungi from *Nothapodytes nimmoniana*. *Fungal Biology*, 120(6–7):873–883.

Bhuvaneswari, S., Madhavan, S., Panneerselvam, A. 2015. Molecular pro-filing and bacteriocin production of endophytic bacteria isolated from *Solanum trilobatum* L. leaves. *International Journal of Current Microbiology and Applied Science* 4:539–546.

Bills, G.F., Dombrowski, A., Peleaz, F. et al. 2002. Recent and future discoveries of pharmacologically acitive metabolites from tropical fungi. In: Watling, R., Frankland, J.C., Ainsworth, A.M. et al. (eds). *Tropical Mycology: Micromycetes*, Vol. 2. CABI Publishing, Wallingford, Oxfordshire, pp. 165–194.

Bills, G.F., Giacobbe, R.A., Lee, S.H., Pelaez, F., Tkacz, J.S. 1992. Tremorgenic mycotoxins paspalitrem A and C from a tropical *Phomopsis*. *Mycological Research* 96:977–983.

Bills, G.F., González-Menéndez, V., Marťın, J. et al. 2012. *Hypoxylon pulicicidum* sp. nov. (Ascomycota, Xylariales), a pantropical insecticide-producing endophyte. *PLoS ONE* 7(10):e46687. doi:10.1371/journal.pone.0046687

Blackwell, M. 2011. The Fungi: 1, 2, 3... 5.1 million species? *American Journal of Botany* 98:426–438.

Borges, K.B., Borges, W.D.S., Pupo, M.T., Bonato, P.S. 2008. Stereoselective analysis of thioridazine-2-sulfoxide and thioridazine-5-sulfoxide: An investigation of *rac*-thioridazine biotransformation by some endophytic fungi. *Journal of Pharmaceutical and Biomedical Analysis* 46(5):945–952. doi:10.1016/j.jpba.2007.05.018

Borges, W.S., Borges, K.B., Bonato, P.S. et al. 2009. Endophytic fungi: Natural products, enzymes and biotransformation reactions. *Current Organic Chemistry* 13(12):1137–1163. doi:10.2174/138527209788921783

Brady, S.F., Singh, M.P., Janso, J.E., Clardy, J. 2000. Cytoskyrins A and B, new BIA active bisanthraquinones isolated from an endophytic fungus. *Organic Letters*, 2(25):4047–4049. https://doi.org/10.1021/ol006681k

Britto, K.C. 1998. *Isolamento e atividade antimicrobiana de actinomicetos endofiticos do feijao (Phaseolus vulgaris L.). Monography.* Brazil: Universidad Federal de Pernambuco, Recife, pernambuco, p. 42.

Brown, A.E., Finlay, R., Ward, J.S. 1987. Antifungal compounds produced by *Epicoccum purpurascens* against soil-borne plant pathogenic fungi. *Soil Biology and Biochemistry* 19:657–664.

Bulla, L.M.C., Polonio, J.C., de Brito Portela-Castro, A.L., Kava, V., Azevedo, J.L., Pamphile, J.A. 2017. Activity of the endophytic fungi *Phlebia* sp. and *Paecilomyces formosus* in decolourisation and the reduction of reactive dyes' cytotoxicity in fish erythrocytes. *Environmental Monitoring Assessment* 189:88. doi:10.1007/s10661-017-5790-0

Bungihan, M.E., Tan, M.A., Kitajima, M., Kogure, N., et al. 2011. Bioactive metabolites of *Diaporthe* sp. P133, an endophytic fungus isolated from *Pandanus amaryllifolius*. *Journal of Natural Medicines*, 65:606–609.

Burkill, H.M. 1985. *The useful plants of West Africa. Families* A-D, vol. 1, Royal Botanic Gardens, Kews: The White Friars Press Limited, United Kingdom.

Bustanussalam, F.R., Septiana, E., Lekatompessy, S.J., Widowati, T., Sukiman, H.I., Simanjuntak, P. 2015. Screening for endophytic fungi from turmeric plant (*Curcuma longa* L.) of Sukabumi and Cibinong with potency as antioxidant compounds producer. *Pakistan Journal of Biological Science* 18:42–45.

Calcul, L., Waterman, C., Ma, W.S. et al. 2013. Screening mangrove endophytic fungi for antimalarial natural products. *Marine Drugs* 11(12):5036–5050. doi:10.3390/md11125036

Campos, F.F., Rosa, L.H., Cota, B.B., Caligiome, R.B., et al. 2008. Leishmanicidal metabolites from *Cochliobolus* sp., an endophytic fungus isolated from *Piptadenia adiantoides* (Fabaceae). *PLOS Neglected Tropical Diseases*, 2(12):e348. https://doi.org/10.1371/journal.pntd.0000348

Cannon, P.F., Simmons, C.M. 2002. Diversity and host preference of leaf endophytic fungi in the Iwokrama Forest Reserve, Guyana. *Mycologia* 94:210–220.

Cantrell, S.A., Casillas-Martinez, L., Molina, M. 2006. Characterization of fungi from hypersaline environments of solar salterns using morphological and molecular techniques. *Mycological Research* 110:962–970.

Capon, R.J., Stewart, M., Ratnayake, R., Lacey, E., Gill, J.H. 2007. Citromycetins and bilains A–C: New aromatic polyketides and diketopiperazines from Australian marine-derived and terrestrial *Penicillium* spp. *Journal Natural Product* 70:1746–1752.

Caro, Y., Anamale, L., Fouillaud, M,, Laurent, P., Petit, T., Dufossé, L. 2012. Natural hydroxyanthraquinoid pigments as potent food grade colorants: An overview. *Natural Product Bioprospecting* 2:174–193.

Carvalho, C.R., Ferreira-D'Silva, A., Wedge, D.E., Cantrell, C.L., Rosa, L.H. 2018. Antifungal activities of cytochalasins produced by *Diaporthe miriciae*, an endo-phytic fungus associated with tropical medicinal plants. *Canadian Journal of Microbiology* 64:835–843. doi:10.1139/cjm-2018-0131

Carvalho, C.R., Gonçalves, V.N., Pereira, C.B. et al. 2012. The diversity, antimicrobial and anticancer activity of endophytic fungi associated with the medicinal plant *Stryphnodendron adstringens* (Mart.) Coville (Fabaceae) from the Brazilian savannah. *Symbiosis* 57:95–107. doi:10.1007/s13199-012-0182-2

Castro, R.A., Quecine, M.C., Lacava, P.T., Batista, B.D., Luvizotto, D.M., Marcon, J., Ferreira, A., Melo, I.S., Azevedo, J.L. 2014. Isolation and enzyme bioprospection of endophytic bacteria associated with plants of Brazilian mangrove ecosystem. *Springer Plus* 3:382.

Chapla, V.M., Zeraik, M.L., Ximenes, V.F. et al. 2014. Bioactive secondary metabolites from *Phomopsis* sp., an endophytic fungus from *Senna spectabilis*. *Molecules* 19:6597–6608. doi:10.3390/molecules19056597

Chen, S., Liu, Z., Li, H., Xia, G., Lu, Y., He, L., Huang, S., She, Z. 2015. β-Resorcylic acid derivatives with α-glucosidase inhibitory activity from *Lasiodiplodia* sp. ZJ-HQ 1, an endophytic fungus in the medicinal plant *Acanthus ilicifolius*. *Phytochemical Letters* 13:141–146.

Chen, X.M., Sang, X.X., Li, S.H., Zhang, S.J., Bai, L.H. 2010. Studies on a chlorogenic acid-prodcuing endophytic fungi isolated from *Eucommia ulmoides* Oliver. *Journal of Industrial Microbiology and Biotechnology* 37(5):447–454.

Chen, X., Schluesener, H.J. 2008. Nanosilver: A product in medical application. *Toxicology Letters*, 176(1):1–12. https://doi.org/10.1016/j.toxlet.2007.10.004

Chen, F., Tan, N., Long, W., Yang, S.K., She, Z.G. 2014. Enhancement of uranium (VI) biosorp-tion by chemically modified marine-derived mangrove endophytic fungus *Fusarium sp.* #ZZF51. *Journal of Radioanalytical and Nuclear Chemistry* 299(1):193–201.

Chen, J., Zhang, L.C., Xing, Y.M. et al. 2013. Diversity and taxonomy of endophytic xylariaceous fungi from medicinal plants of *Dendrobium* (Orchidaceae). *PLoS ONE* 8(3):e58268. doi:10.1371/journal.pone.0058268

Cheplick, G.P. 2004. Symbiotic fungi and clonal plant physiology. *The New Phytologist* 164(3):413–415.

Choi, J.N., Kim, J., Ponnusamy, K., Lim, C., Kim, J.G., Muthaiya, M.J., Lee, C.H. 2013a. Identification of a new phomoxanthone antibiotic from *Phomopsis longicolla* and its antimicrobial correlation with other metabolites during fermentation. *The Journal of Anibiotics*, 66:231–233. https://doi.org/10.1038/ja.2012.105

Choi, J.N., Kim, J., Ponnusamy, K., Lim, C., Kim, J.G., Muthaiya, M.J., Lee, C.H. 2013b. Metabolic changes of Phomopsis longicolla fermentation and its effect on antimicrobial activity against *Xanthomonas oryzae*. *Journal of Microbial Biotechnology*, 23(2):177–183. http://dx.doi.org/10.4014/jmb.1210.10020

Chomcheon, P., Sriubolmas, N., Wiyakrutta, S. et al. 2006. Cyclopentenones, scaffolds for organic syntheses produced by the endophytic fungus mitosporic *Dothideomycete* sp. LRUB20. *Journal of Natural Product* 69(9):1351–1353. doi:10.1021/np060148h

Chow, Y.Y., Ting, A.S.Y. 2015. Endophytic L-asparaginase-producing fungi from plants associated with anticancer properties. *Journal of Advanced Research* 6(6):869–876.

Chowdhary, K., Kaushik, N. 2015. Fungal endophyte diversity and bioactivity in the Indian medicinal plant *Ocimum sanctum* Linn. *PLoS ONE* 10:e0141444. doi:10.1371/journal.pone.0141444

Christian, N., Sullvan, C., VIsser, N.D., Clay, K. 2016. Plant host and geographic location drive endophyte community composition in the face of perturbation. *Microbial Ecology* 72:621–632.

Cimmino, A., Andolfi, A., Berestetskiy, A., Evidente, A. 2008. Production of phytotoxins by *Phoma exigua* var. *exigua*, a potential mycoherbicide against perennial thistles. *Journal of Agriculture and Food Chemistry* 56:6304–6309. doi:10.1021/jf8004178

Corpe, W.A., Rheem, S. 1989. Ecology of the methylotrophic bacteria on living leaf surfaces. *FEMS Microbiology Ecology* 62:243–250.

Corrêa, R.C.G., Rhoden, S.A., Mota, T.R., Azevedo, J.L., Pamphile, J.A., de Souza, C.G.M., de Moraes Polizeli, M.L.T., Bracht, A., Peralta, R.M. 2014. Endophytic fungi: Expanding the arsenal of industrial enzyme producers. *Journal of Industrial Microbiology and Biotechnology* 41:1467–1478.

Cummings, N.J., Ambrose, A., Braithwaite, M., Bissett, J., Roslan, H.A., Abdullah, J., Stewart, A., Agbayani, F.V., Steyaert, J., Hill, R.A. 2016. Diversity of root-endophytic *Trichoderma* from Malaysian Borneo. *Mycological Progress* 15:50. doi:10.1007/s11557-016-1192-x

da Costa Maia, N., da Costa Souza, P.N., Godinho, B.T.V., Moreira, S.I., de Abreu, L.M., Jank, L., Cardoso, P.G. 2018. Fungal endophytes of *Panicum maximum* and *Pennisetum purpureum*: Isolation, identification, and determination of antifungal potential. *Brazilian Journal of Animal Science*. Revista Brasileira de Zootecnia 47:e20170183. doi:10.1590/rbz4720170183

Dai, C., Tian, L., Zhao, Y. et al. 2010. Degradation of pheanthrene by the endophytic fungus *Ceratobasium stevensii* found in *Bishofia polycarpa*. *Biodegradation* 21:245. doi:10.1007/s10532-009-9297-4

Dai, J.Q., Krohn, K., Florke, U. et al. 2005. Novel highly substituted biaryl ethers, phomosines D-G, isolated from the endophytic fungus *Phomopsis* sp. from *Adenocarpus foliolosus*. *European Journal of Organic Chemistry* 2005(1). doi:10.1002/ejoc.200500471

Daisy, B.H., Strobel, G.A., Castillo, U., Ezra, D., Sears, J., Weaver, D.K., Runyon, J.B. 2002. Naphthalene, an insect repellent, is produce by *Muscodor vitigenus*, a new endophytic fungus. *Microbiology* 148(11):3737–3741.

Dalee, A.D., Mukhurah, S., Sali, K., Hayeeyusoh, N., Hajiwangoh, Z., Salaeh, P. 2015. Antimicrobial substances from endophytic fungi in tamarind (*Tamarindus indica*, Linn), Malay apple (*Eugenia malaccensis*, Linn), rambutan (*Nephelium lappaceum*), and Indian mulberry (*Morinda citrifolia*, Linn). In: *Proceeding of International Conference on Research, Implementation and Education of Mathematics and Sciences 2015*. Faculty of Mathematics and Natural Sciences Yogyakarta State University.

Davis, E.C., Franklin, J.B., Shaw, A.J., Vilgalys, R. 2003. Endophytic *Xylaria* (Xylariaceae) among liverworts and angiosperms: Phylogenetics, distribution and symbiosis. *American Journal of Botany*, 90(11):1661–1667. https://doi.org/10.3732/ajb.90.11.1661

Davis, E.C., Shaw, A.J. 2008. Biogeographic and phylogenetic patterns in diversity of liverwort-associated endophytes. *American Journal of Botany* 95:914–924.

Deng, Z.J., Zhang, R.D., Shi, Y., Hu, L., Tan, H.M., Cao, L.X. 2014. Characterization of Cd-, Pb-, Zn-resistant endophytic *Lasiodiplodia* sp. MXSF31 from metal accumulating *Portulaca oleracea* and its potential in promoting the growth of rape in metal-contaminated soils. *Environmental Science Pollution Research* 21:2346–2357. doi:10.1007/s11356-013-2163-2

Dettrakul, S., Kittakoop, P., Isaka, M., Nopichai, S., Suyarnsestakorn, C., Tanticharoen, Thebtaranontha, M.Y. 2003. Antimycobacterial pimarane diterpenes from the fungus *Diaporthe* sp. *Bioorganic and Medicinal Chemistry Letters*, 13(7):1253–1255. https://doi.org/10.1016/S0960-894X(03)00111-2

Ding, B., Yuan, J., Huang, X.S. et al. 2013. New dimeric members of the phomoxanthone family: Phomolactonexanthones A, B and deacetylphomoxanthone C isolated from the fungus *Phomopsis sp. Marine Drugs* 11(12):4961–4972. doi:10.3390/md11124961

Dissanayake, A.J., Hyde, K.D. 2015. Morphological and molecular characterization of *Diaporthe* species associated with grapevine trunk disease in China. *Fungal Biology*, 119(5):283–294. https://doi.org/10.1016/j.funbio.2014.11.003

Dnyaneshwar, G., Devi, P., Supriya, T., Naik, C.G., Parameswaran, P.S. 2002. Fungal metabolites: Tetrahydroauroglaucin and isodihydroauroglaucin from the marine fungus, *Eurotium sp.* In: Sree, A., Rao, Y.R., Nanda, B., Misra, V.N. (eds). *Proceedings of National Conference on Utilization of Bioresources—NATCUB-2002 October 24–25*. Bhubaneswar: Regional Research Laboratory. 2002:453–457.

Doley, P., Jha, D.K. 2016. Antimicrobial activity of bacterial endophytes from medicinal endemic plant *Garcinia lancifolia* Roxb. *Annals of Plant Science* 4:1243–1247.

Doumbou, C.L., Akimov, V., Beaulieu, C. 1998. Selection and characterization of microorganisms utilizing thaxtomin A, a phytotoxin produced by *Streptomyces scabies*. *Applied Environmental Microbiology* 64:4313–4316.

Druzhinina, I.S., Seidl-Seiboth, V., Herrera-Estrella, A., Horowitz, B.A., et al. 2011. *Trichoderma*: The genomics of opportunistic success. *Nature Reviews Microbiology*, 9:749–759. https://doi.org/10.1038/nrmicro2637

Du, L., King, J.B., Morrow, B.H., Shen, J.K., Miller, A.N., Cichewicz, R.H. 2012. Diarylcyclopentendione metabolite obtained from a *Preussia typharum* isolate procured using an unconventional cultivation approach. *Journal of Natural Products*, 75(10):1819–1823. https://doi.org/10.1021/np300473h

Dufosse, L., Fouillaud, M., Caro, Y., Mapari, S.A.S., Sutthiwong, N. 2014. Filamentous fungi are large-scale producers of pigments and colorants for the food industry. *Current Opinion in Biotechnology* 26:56–61. https://doi.org/10.1016/j.copbio.2013.09.007

Duong, L.M., Jeewon, R., Lumyong, S., Hyde, K.D. 2006. DGGE coupled with ribosomal DNA phylogenies reveal uncharacterized fungal phylotypes on living leaves of *Magnolia liliifera. Fungal Diversity* 23:121–138.

Eaton, C.J., Cox, M.P., Scott, B. 2011. What triggers grass endophytes to switch from mutualism to pathogenism? *Plant Science* 180(2):190–195.

Egamberdieva, D., Allah, E.F.A. 2017. Impact of soil salinity on plant growth-promoting and biological control abilities of root associated bacteria. *Saudi Journal of Biological Sciences* 24(7):1601–1608.

Elad, Y. Mechanisms involved in the biological control of *Botrytis cinerea* incited diseases. *European Journal of Plant Pathology*, 102:719–732. https://doi.org/10.1007/BF01877146

El-Gendy, M.M., Al-Zahrani, S.H., El-Bondkly, A.M. 2017. Construction of potent recombinant strain through intergeneric protoplast fusion in endophytic fungi for anticancerous enzymes production using rice straw. *Applied Biochemiestry and Biotechnology*, 183:30–50. https://doi.org/10.1007/s12010-017-2429-0

Elmoslamy, S.H., Elkady, M.F., Rezk, A.H., Abdelfattah, Y.R. 2017. Applying taguchi design and large-scale strategy for mycosynthesis of nano-silver from endophytic *Trichoderma harzianum* SYA.F4 and its application against phytopathogens. *Science Reports*, 7:45297. https://doi.org/10.1038/srep45297

Elsasser, B., Krohn, K., Florke, U. et al. 2005. X-ray structure determination, absolute configuration and biological activity of Phomaxanthone A. *European Jounral of Organic Chemistry* 2005(21). doi:10.1002/ejoc.200500265

Erbert, C., Lopes, A.A., Yokoya, N.S., Furtado, Niege A.J.C., Conti, R., Pupo, M.T., Lopes, J.L.C., Debonsi, H.M. 2012. Antibacterial compound from the endophytic fungus *Phomopsis longicolla* isolated from the tropical red seaweed *Bostrychia radicans. Botanica Marina* 55:435–440. doi:10.1515/bot-2011-0023

Ernawati, M., Solihin, D.D., Lestari, Y. 2016. Community structures of endophytic actinobacteria from medicinal plant *Centella asiatica* L. urban-based on metagenomic approach. *International Journal of Pharmacological Science* 8:292–297.

Evidente, A., Rodeva, R., Andolfi, A. et al. 2011. Phytotoxic polyketides produced by *Phomopsis foeniculi*, a strain isolated from diseased Bulgarian fennel. *European Journal of Plant Pathology* 130:173. doi:10.1007/s10658-011-9743-0

Ezra, D., Hess, W.M., Strobel, G.A. 2004. New endophytic isolates of *Muscodor albus*, a volatile-antibiotic-producing fungus. *Microbiology* 150:4023–4031.

Faeth, S.H., Saari, S. 2012. Fungal grass endophytes and arthropod communities: Lessons from plant defence theory and multitrophic interactions. *Fungal Ecology* 5:364–371.

Fatima, K., Imran, A., Amin, I., Khan, Q.M., Afzal, M. 2016. Plant species affect colonization patterns and metabolic activity of associated endophytes during phytoremediation of crude oil-contaminated soil. *Environmental Science Pollution Research* 23:6188–6196. doi:10.1007/s11356-015-5845-0

Feng, Y., Shao, Y., Chen, F. 2012. *Monascus* pigments. *Applied Microbiology and Biotechnology* 96:1421–1440.

Fischer, M.S., Rodriguez, R.J. 2013. Fungal endophytes of invasive *Phagramites australis* populations vary in species composition and fungicide susceptibility. *Symbiosis* 61:55–62.

Fisher, P.J., Petrini, L.E., Sutton, B.C. 1995. A study of fungal endophytes from leaves, stem and roots of *Gynoxis oleifolia*, Muchler (Compositae) from Ecuador. *Nova Hedwigia* 60:589–594.

Friesen, M.L. 2013. Microbially mediated plant functional traits. In: Bruijn, F.J. (Ed.), *Molecular Microbial Ecology of the Rhizosphere*, Vol. 1. John Wiley & Sons, Inc., Hoboken, NJ, pp. 87–102.

Frisvad, J.C., Andersen, B., Thrane, U. 2008. The use of secondary metabolite profiling in chemotaxonomy of filamentous fungi. *Mycological Research* 112:231–240.

Fuentes-Ramírez, L.E., Caballero-Mellado, J., Sepúlveda, J., Martínez-Romero, E. 1999. Colonization of sugarcane by *Acetobacter diazotrophicus* is inhibited by high N-fertilization. *FEMS Microbiology Ecology* 29:117–128.

Fuqua, W.C., Winans, S.C., Greenberg, E.P. 1994. Quorum sensing in bacteria: The LuxR-LuxI family of cell density-responsive transcriptional regulators. *Journal of Bacteriology* 176:269–2675.

Gallo, M.B., Chagas, F.O., Almeida, M.O. et al. 2009. Endophytic fungi found in association with *Smallanthus sonchifolius* (Asteraceae) as resourceful producers of cytotoxic bioactive natural products. *Journal of Basic Microbiology* 49(2). doi:10.1002/jobm.200800093

Gange, A.C., Eschen, R., Wearn, J.A., Thawer, A., Sutton, B.C. 2012. Differential effects of foliar endophytic fungi on insect herbivores attacking a herbaceous plant. *Oecologia* 168:1023. doi:10.1007/s00442-011-2151-5

Gao, Y., Zhao, J.T., Zu, Y.G. et al. 2011. Characterization of five fungal endophytes producing cajaninstilbene acid isolated from pigeon pea (*Cajanus cajan* (L.) Millsp.). *PLoS ONE* 6(11):e27589. doi:10.1371/journal.pone.0027589

Gazis, R., Chaverri, P. 2010. Diversity of fungal endophytes in leaves and stems of wild rubber trees (*Hevea brasiliensis*) in Peru. *Fungal Ecology* 3:240–254.

Gazis, R., Miadlikowska, J., Lutzoni, F., Arnold, A.E., Chaverri, P. 2012. Culture-based study of endophytes associated with rubber trees in Peru reveals a new class of Pezizomycotina: Xylonomycetes. *Molecular Phylogenetics and Evolution* 65:294–304.

Gerhardt, K.E., Huang, X.D., Glick, B.R., Greenberg, B.M. 2009. Phytoremediation and rhizoremediation of organic soil contaminants: Potential and challenges. *Plant Science* 176:20–30.

Germaine, K.J., Keogh, E., Ryan, D., Dowling, D.N. 2009. Bacterial endophyte-mediated naphthalene phytoprotection and phytoremediation. *FEMS Microbiology Letters* 296(2):226–234. doi:10.1111/j.1574-6968.2009.01637.x

Gessler, N.N., Egorova, A.S., Belozerskaya, T.A. 2013. Fungal anthraquinones. *Applied Biochemistry and Microbiology* 49:109–123.

Glick, B.R. 2005. Modulation of plant ethylene levels by the bacterial enzyme ACC deaminase. *FEMS Microbiology Letters* 251:1–7.

Golinska, P., Wypij, M., Rathod, D. et al. 2016. Synthesis of silver nanoparticles from two acidophilic strains of *Pilimelia columellifera* ubsp. *pallida* and their antibacterial activities. *Journal of Basic Microbiology* 56(5). doi:10.1002/jobm.201500516

Gong, B., Chen, Y.P., Zhang, H., Zheng, X., Zhang, Y.Q., Fang, H.Y., Zhong, Q.P., Zhang, C.X. 2014. Isolation, characterization and anti-multiple drug resistant (MDR) bacterial activity of endophytic fungi isolated from the mangrove plant, *Aegiceras corniculatum*. *Tropical Journal of Pharmaceutical Research* 13(4):593–599. doi:10.4314/tjpr.v13i4.16

Gong, B., Liu, G.X., Liao, R., Song, J.J., Zhang, H. 2017. Endophytic fungus Purpureocillium sp. A5 protect mangrove plant *Kandelia candel* under copper stress. *Brazilian Journal of Microbiology* 48:530–536.

González-Rodríguez, M.L., Mouram, I., Ma, C.B., Villasmil, S., Rabasco, A.M. 2012. Applying the taguchi method to optimize sumatriptan succinate niosomes as drug carriers for skin delivery. *Journal of Pharmaceutical Sciences*, 101(10):3845–3363. https://doi.org/10.1002/jps.23252

Gomes, R.R., Glienke, C., Videira, S.I.R., Lombard, L., Groenewald, J.Z., Crous, P.W. 2013. *Diaporthe*: A genus of endophytic, saprobic and plant pathogenic fungi. *Persoonia*, 31:1–41. https://doi.org/10.3767/003158513X666844

Gouda, S., Das, G., Sen, S.K., Shin, H.S., Patra, J.K. 2016. Endophytes: A treasure house of bioactive compounds of medicinal importance. *Frontiers in Microbiology*. doi:10.3389/fmicb.2016.01538

Govindarajan, M., Balandreau, J., Kwon, S.W., Weon, H.Y., Lakshminarasimhan, C. 2008. Effects of the inoculation of *Burkholderia vietnamensis* and related endophytic diazotrophic bacteria on grain yield of rice. *Microbial Ecology* 55:21–37.

Govindarajan, M., Kwon, S.W., Weon, H.Y. 2007. Isolation, molecular characterization and growth-promoting activities of endophytic sugarcane diazotroph *Klebsiella* sp. GR9. *World Journal of Microbiology and Biotechnology* 23(7): 997–1006.

Gracida-Rodriguez, J., Gomez-Valadez, A., Tovar-Jimenez, X., Amaro-Reyes, A., Arana-Cuenca, A., Zamudio-Perez, E. 2017. Optimization of the biosynthesis of napthoquinones by endophytic fungi isolated from *Ferocactus latispinus*. *Biologia* 72(12):1416–1421. https://doi.org/10.1515/biolog-2017-0177

Griffin, E.A., Carson, W.P. 2015. The ecology and natural history of foliar bacteria with a focus on tropical forests and agroecosystems. *The Botanical Review* 81:105–149.

Griffin, M.A., Spakowicz, D.J., Gianoulis, T.A., Strobel, S.A. 2010. Volatile organic compound production by organisms in the genus *Ascocoryne* and a re-evaluation of mycodiesel production by NRRL 50072. *Microbiology* 156(12):3814–3829. doi:10.1099/mic.0.041327-0

Gunasekera, T.S., Sundin, G.W. 2006. Role of nucleotide excision repair and photoreactivation in the solar UVB radiation survival of *Pseudomonas syringae* pv. *syringae* B728a. *Journal of Applied Microbiology* 100:1073–1083.

Gupta, R., Beg, Q. Lorenz, P. 2002. Bacterial alkaline proteases: Molecular approaches and industrial applications. *Applied Microbiology and Biotechnology*, 59:15–32. https://doi.org/10.1007/s00253-002-0975-y

Guo, L.D., Hyde, K.D., Liew, E.C.Y. 2001. Detection and taxonomic placement of endophytic fungi within frond tissues of *Livistona chinensis* based on rDNA sequences. *Molecular Phylogenetics and Evolution* 19:1–13.

Haas, D., Défago, G. 2005. Biological control of soil-borne pathogens by fluorescent pseudomonads. *Nature Review Microbiology* 3:307–319.

Hallmann, J., Quadt-Hallman, A., Mahafee, W.F., Kloepper, J.W. 1997. Bacterial endophytes in agricultural crops. *Canadian Journal of Microbiology* 43:895–914.

Hanada, R.E., Souza, T.J., Pomella, A.W.V., Hebbar, K.P., Rereira, J.O., Ismaiel, A., Samuels, G.J. 2008. *Trichoderma martiale* sp. nov., a new endophyte from sapwood of *Theobroma cacao* with a potential for biological control. *Mycological Research*, 112(11):1335–1343. https://doi.org/10.1016/j.mycres.2008.06.022

Hardoim, P.R., van Overbeek, L.S., Berg, G. et al. 2015. The hidden world within plants: Ecological and evolutionary considerations for defining fundctioning og microbial endophytes. *Microbiology and Molecular Biology Reviews*. doi:10.1128/MMBR.00050-14

Harman, G.E. 2000. Myths and dogmas of biocontrol changes in perceptions derived from research on *Trichoderma harzinum* T-22. *Plant Disease* 84:377–393.

Harman, G.E. 2011. Multifunctional fungal plant symbionts: New tools to enhance plant growth and productivity. *New Phytologist* 189:647–649.

Harman, G.E., Howell, C.R., Viterbo, A., Chet, I., Lorito, M. 2004. *Trichoderma* species—Opportunistic, avirulent plant symbionts. *Nature Review Microbiology* 2:43–56.

Harper, J.K., Arif, A.M., Ford, E.J. et al. 2003. Pestacin: A 1,3-dihydro isoben-zofuran from *Pestalotiopsis microspora* possessing antioxidant and antimycotic activities. *Tetrahedron* 59:2471–2476.

Harrison, L., Teplow, D.B., Rinaldi, M., Strobel, G. 1991. Pseudomycins, a family of novel peptides from *Pseudomonas synringae* possessing broad-spectrum antifungal activity. *Microbiology* 137(12):2857–2865.

Haruna, E., Zin, N.M., Kerfahi, D., Adams, J.M. 2018. Extensive overlap of tropical rainforest bacterial endophytes between soil, plant parts and plant species. *Microbial Ecology* 75:88–103.

Hawksworth, D.L. 1991. The fungal dimension of biodiversity: Magnitude, significance, and conservation. *Mycological Research* 95:641–655.

Hawksworth, D.L. 2001. The magnitude of fungal biodiversity: The 1.5 million species estimate revisited. *Mycological Research* 105:1422–1435.

Hazalin, N.A.M.N., Ramasamy, K., Lim, S.M., Wahab, I.A., Cole, A.L.J., Majeed, A.B.A. 2009. Cytotoxic and antibacterial activities of endophytic fungi isolated from plants at the National Park, Pahang, Malaysia. *BMC Complementary and Alternative Medicine* 9:46. doi:10.1186/1472-6882-9-46

Helander, M.L., Neuvonen, S., Sieber, T., Petrini, O. 1993. Simulated acid rain affects birch leaf endophyte populations. *Microbial Ecology* 26:227–234.

Hermosa, R., Viterbo, A., Chet, I., Monte, E. 2012. Plant-beneficial effects of Trichoderma and of its genes. *Microbiology*, 158(1):17–25. doi:10.1099/mic.0.052274-0

Herre, E.A., Mejia, L.C., Kyllo, D.A., Rojas, E., Maynard, Z., Butler, A., vanBael, S.A. 2007. Ecological implications of anti-pathogen effects of tropical fungal endophytes and mycorrhizae. *Ecology* 88:550–558.

Herrera, C.M., De Vega, C., Canto, A., Pozo, M.I. 2009. Yeasts in floral nectar: A quantitative survey. *Annals in Botany* 103:1415–1423.

Higginbotham, S.J., Arnold, A.E., Ibanez, A., Spadafora, C., Coley, P.D., Kursar, T.A. 2013. Bioactivity of fungal endophytes as a function of endophyte taxonomy and the taxonomy and distribution of their host plants. *PLoS ONE* 8(9):e73192.

Higgins, K.L., Arnold, A.E., Coley, P.D., Kursar, T.A. 2014. Communities of fungal endophytes in tropical forest grasses: Highly diverse host- and habitat generalists characterized by strong spatial structure. *Fungal Ecology* 8:1–11.

Higgins, K.L., Coley, P.D., Kursar, T.A., Arnold, A.E. 2011. Culturing and direct PCR suggest prevalent host generalism among diverse fungal endophytes of tropical forest grasses. *Mycologia* 103:247–260.

Ho, Y.N., Chiang, H.M., Chao, C.P., Su, C.C., Hsu, H.F., Guo, C.T., Hsieh, J.L., Huang, C.C. 2015. *In planta* biocontrol of soilborne Fusarium wilt of banana through a plant endophytic bacterium, *Burkholderia cenocepacia* 869T2. *Plant Soil* 387:295–306. doi:10.1007/s11104-014-2297-0

Holmes, K.A., Schroers, H., Thomas, S.E., et al. 2004. Taxonomy and biocontrol potential of a new species of *Trichoderma* from the Amazon basin of South America. *Mycological Progress*, 3:199–210. https://doi.org/10.1007/s11557-006-0090-z

Horn, W.S., Simmonds, M.S.J., Schwatz, R.E., Blaney, W.M. 1995. Phomopsichalasin, a novel antimicrobial agent from an endophytic *Phomopsis* sp. *Tetrahedron* 51:3969–3978.

Howell, C.R. 2003. Mechanisms employed by *Trichoderma* species in the biological control of plant diseases: The history and evolution of current concepts. *Plant Disease* 87:4–10.

Hoyos-Carvajal, L., Bissett, J. 2011. Biodiversity of *Trichoderma* in neotropics. In: Grillo, O., Venora, G. (eds). *The Dynamical Processes of Biodiversity: Case Studies of Evolution and Spatial Distribution.* InTech, Rijeka, pp. 303–320.

Huang, C.H., Pan, J.H., Chen, B., Yu, M., Huang, H.B., Zhu, X., Lu, Y.J., She, Z.G., Lin, Y.C. 2011. Three bianthraquinone derivatives from the mangrove endophytic fungus *Alternaria* sp. ZJ9-6B from the South China Sea. *Marine Drugs* 9:832–843.

Huang, W.Y., Cai, Y.Z., Hyde, K.D., Corke, H., Sun, M. 2008. Biodiversity of endophytic fungi associated with 29 traditional Chinese medicinal plants. *Fungal Diversity* 33:61–75.

Huang, Y., Wang, J., Li, G., Zheng, Z., Su, W. 2001. Antitumor and antifungal activities in endophytic fungi isolated from pharmaceutical plants *Taxus mairei*, *Cephalataxus fortunei* and *Torreya grandis*. *FEMS Immunology Medical Microbiology* 31(2):163–167. doi:10.1111/j.1574-695X.2001.tb00513.x

Huang, Y.L., Devan, M.M.N., U'Ren, J.M. et al. 2015. Pervasive effects of wildfire on foliar endophyte communities in montane forest trees. *Microbial Ecology* 71:452–468.

Huang, Z., Yang, R., Guo, Z., She, Z., Lin, Y. 2010. New anthraquinone derivative produced by cultivation of mangrove endophytic fungus *Fusarium* sp. ZZF60 from the South China Sea. *Chinese Journal of Applied Chemisty* 27:394–397.

Huo, W., Zhuang, C.H., Cao, Y., Pu, Meng, et al. Paclobutrazol and plant-growth promoting bacterial endophyte *Pantoea* sp. enhance copper tolerance of guinea grass (*Panicum maximum*) in hydroponic culture. *Acta Physiologiae Plantarum*, 34:139–150. https://doi.org/10.1007/s11738-011-0812-y

Hyde, K.D., Soytong, K. 2008. The fungal endophyte dilemma. *Fungal Diversity* 33:163–173.

Idris, R., Trifonova, R., Puschenreiter, M., Welzel, W.W., Seissitsch, A. 2004. Bacterial communities associated with flowering plants of the Ni hyperaccumulator *Thlaspi goesingense*. *Applied Environmental Microbiology* 70:2667–2677.

Ikawa, M., McGratten, C.J., Burge, W.R. et al. 1978. Epirodin, a polyene antibiotic from the mold *Epipoccum nigrum*. *Journal of Antibiotics* 31:159–161.

Isaka, M., Jaturapat, A., Rukseree, K., Danwisetkanjana, K., Tanticharoen, M., Thebtaranonth, Y. 2001. Phomoxanthones A and B, novel xanthone dimers from the endophytic fungus *Phomopsis species. Journal of Natural Products*, 64(8):1015–1018. https://doi.org/10.1021/np010006h

Jacobs, J.L., Carroll, T.L., Sundin, G.W. 2005. The role of pigmentation, ultraviolet radiation tolerance, and leaf colonization strategies in the epiphytic survival of phyllosphere bacteria. *Microbial Ecology* 49:104–113.

Jallow, M.F.A., Dugassa-Gobena, D., Vidal, S., 2004. Indirect interaction between an unspecialized endophytic fungus and a polyphagous moth. *Basic and Applied Ecology* 5:183–191.

James, E.K., Olivares, F.L. 1998. Infection and colonization of sugarcane and other graminaceous plants by endophytic diazotrophs. *Critical Reviews in Plant Sciences* 17:77–119.

Jewett, T.J., Fischer, E.R., Mead, D.J., Hackstadt, T. 2006. Chlamydial TARP is a bacterial nucleator of actin. *Proceedings of the National Academy of Sciences USA* 103:15599–15604.

Jimenez-Romero, C., Ortega-Barria, E., Arnold, A.E., Cubilla-Rios, L. 2008. Activity against *Plasmodium falciparum* of lactones isolated from the endophytic fungus *Xylaria* sp. *Pharmaceutical Biology* 46:1–4.

Jobanputra, A.H., Patil, G.D., Sayyed, R.Z, Chaudhari, A.B., Chincholkar, S.B. 2003. Microbial transformation of rifamycin: A novel approach to rifamycin derivatives. *Indian Journal of Biotechnology* 2(3):370–377. http://nopr.niscair.res.in/handle/123456789/11321

Johnson, G.I., Mead, A.J., Cooke, A.W., Dean, J.R. 1992. Mango stem and rot pathogens: Fruit infection by endophytic colonization of the inflorescence and pedicel. *Annals of Applied Biology* 120:225–234.

Jumpponen, A., Jones, K.L. 2009. Massively parallel 454 sequencing indicates hyperdiverse fungal communities in temperate *Quercus macrocarpa* phyllosphere. *New Phytologist* 184:438–448.

Kamdem, R.S.T., Pascal, W., Rehberg, N., van Geelen, L., Höfert, S.P., Knedel, T.O., Janiak, C., Sureechatchaiyan, P., Kassack, M.U., Lin, W.H., Kalscheuer, R., Liu, Z., Proksch, P. 2018a. Metabolites from the endophytic fungus *Cylindrocarpon* sp. isolated from tropical plant *Sapium ellipticum. Fitoterapia* 128:175–179.

Kamdem, R.S.T., Wang, H., Wafo, P., Ebrahim, W., Özkayaa, F.C., Makhloufi, G., Janiak, C., Sureechatchaiyan, P., Kassack, M.U., Lin, W., Liu, Z., Proksch, P. 2018b. Induction of new metabolites from the endophytic fungus *Bionectria* sp. through bacterial co-culture. *Fitoterapia* 124:132–136.

Kaneko, R., Kaneko, S. 2004. The effect of bagging branches on levels of endophytic fungal infection in Japanese beech leaves. *Forest Pathology* 34:65–78.

Kaul, S., Ahmed, M., Sharma, T., Dhar, M.K. 2014. Unlocking the myriad benefits of endophytes: An Overview. In Kharwar, R.N., Upadhyay, R.S., Dubey, N.K., Raghuwanshi, R. (Eds.), *Microbial Diversity and Biotechnology in Food Security.* Springer India, New Delhi, pp. 41–57. https://doi.org/10.1007/978-81-322-1801-2_4

Kaushik, N.K., Murali, T.S., Sahal, D., Suryanarayanan, T.S. 2014. A search for antiplasmodial metabolites among fungal endophytes of terrestrial and marine plants of southern India. *Acta Parasitologica* 59(4):745–757. doi:10.2478/s11686-014-0307-2

Kei, H., Nina, G., Soledad, S., Thomas, N.R., Stéphane, H., Barbara, K., Ulla, N., Diana, R., Marcel, B., O'Connell, R.J. 2016. Root endophyte *Colletotrichum tofieldiae* confers plant fitness benefits that are phosphate status dependent. *Cell* 165(2):464–474. doi:10.1016/j.cell.2016.02.028

Khan, A.L., Hamayun, M., Kang, S.M., Kim, Y.H., Jung, H.Y., Lee, J.H., Lee, I.J. 2012. Endophytic fungal association via gibberellins and indole acetic acid can improve plant growth under abiotic stress: An example of *Paecilomyces formosus* LHL10. *BMC Microbiology* 12(1):3. doi:10.1186/1471-2180-12-3

Khan, A.R., Ullah, I., Waqas, M., Park, G.S., Khan, A.L., Hong, S.J., Ullah, R., Jung, B.K., Park, C.E., Ur-Rehman, S. 2017. Host plant growth promotion and cadmium detoxification in *Solanum nigrum*, mediated by endophytic fungi. *Ecotoxicology and Environmental Safety* 136(2017):180–188. doi:10.1016/j. ecoenv.2016.03.014

Khiralla, A., Mohamed, I., Thomas, J., Mignard, B., Spina, R., Yagi, S., Laurain-Mattar, D. 2015. A pilot study of antioxidant potential of endophytic fungi from some Sudanese medicinal plants. *Asian Pacific Journal of Tropical Medicine* 8(9):701–704. doi:10.1016/j.apjtm.2015.07.032

Kim, J.C., Choi, G.J., Park, J.H., Kim, H.T., Cho, K.Y. 2001. Activity against plant pathogenic fungi of phomalactone isolated from *Nigrospora spaherica*. *Pest Management Science* 57(6). doi:10.1002/ps.318

Kim, J.J., Sundin, G.W. 2001. Construction and analysis of photolease mutants of *Pseudomonas aeruginosa* and *Pseudomonas syringae*: Contribution of photoreactivation, nucleotide excision repair, and mutagenic DNA repair to cell survival and mutability following exposure to UV-B radiation. *Applied and Environmental Microbiology* 67:1405–1411.

Kimura, N. 2006. Metagenomics: Access to unculturable microbes in the environment. *Microbes and Environments* 21:201–215.

King, T.J., Roberts, J.C., Thompson, D.J. 1970. The structure of purpurogenone, a metabolite of Penicillium purpurogenum stoll: An X-ray study. *Journal of the Chemical Society D: Chemical Communications* 22:1499a. doi:10.1039/ C2970001499A

Klaiklay, S., Rukachaisirikul, V., Phongpaichit, S. et al. 2012. Anthraquinone derivatives from the mangrove-derived fungus *Phomopsis* sp. PSU-MA214. *Phytochemistry Letters* 5(4):738–742.

Kobayashi, H., Meguro, S., Yoshimoto, T., Namikoshi, M. 2003. Absolute structure, biosynthesis, and anti-microtubule activity of phomopsidin, isolated from a marine-derived fungus *Phomopsis sp. Tetrahedron* 59(4):455–459.

Krings, M., Taylor, T.N., Dotzler, N. 2012. Fungal endophytes as a driving force in land plant evolution: Evidence from the fossil record. In: Southworth (Ed). *Biocomplexity of Plant-Fungal Interactions*. John Wiley & Sons, Ames, IA, pp. 5–27.

Krohn, K., Michel, A., Romer, E. et al. 1995. Biologically active metabolites from fungi 61; Phomosines A-C three new biaryl ethers from *Phomopsis* sp. *Natural Product Letters* 6(4). doi:10.1080/10575639508043176

Kruasuwan, W., Thamchaipenet, A. 2016. Diversity of culturable plant growth-promoting bacterial endophytes associated with sugarcane roots and their effect of growth by co-inoculation of diazotrophs and actinomycetes. *Journal of Plant Growth Regulators* 35:1074–1087. doi:10.1007/s00344-016-9604-3

Kuffner, M., DeMaria, S., Puschenreiter, M., Fallmann, K., Wieshammer, G., et al. 2010. Culturable bacteria from Zn- and Cd-accumulating *Salix caprea* with differential effects on plant growth and heavy metal availability. *Journal of Applied Microbiology*, 108:1471–1484. https://doi.org/10.1111/j.1365-2672.2010.04670.x

Kuklinsky-Sobral, J., Araujo, W.L., Mendes, R. et al. 2004. Isolation and characterization of soybean-associated bacteria and their potential for plant growth promotion. *Environmental Microbiology* 6(12). doi:10.1111/j.1462-2920.2004.00658.x

Kumar, A., Ahmad, A. 2013. Biotransformation of vinblastine to vincristine by the endophytic fungus Fusarium oxysporum isolated from Catharanthus roseus. *Biocatalysis and Biotransformation* 31(2):89–93.

Kumar, S., Kaushik, N., Edrada-Ebel, R., Ebel, R., Proksch, P. 2011. Isolation, characterization, and bioactivity of endophytic fungi of *Tylophora indica*. *World Journal of Microbiology and Biotechnology* 27(3):571–577. doi:10.1007/s11274-010-0492-6

Kumara, P.M., Zuehlke, S., Priti, V. et al. 2012. *Fusarium proleferatum*, an endophytic fungus from *Dysoxylum binectariferum* Hook.f, produces rohitukine, a chromane alkaloid possessing anticancer activity. *Antonie van Leeuwenhoek* 101:323–329.

Kumaresan, V., Suryanarayanan, T.S., Johnson, J.A. 2002. Ecology of mangrove endophytes: Fungal diversity research series 7. In: Hyde, K.D. (ed). *Fungi in Marine Environments*. Fungal Diversity Press, Hong Kong, pp. 145–166.

Lebeau, T., Braud, A., Jézéquel, K. 2008. Performance of bioaugmentation-assisted phytoextraction applied to metal contaminated soils: A review. *Environmental Pollution* 153:497–522.

Lee, J.C., Strobel, G.A, Lobkovsky, E., Clardy, J. 1996. Torreyanic acid: A selectively cytotoxic quinone dimer from the endophytic fungus Pestlotiopsis microspora. *Journal of Organic Chemistry* 61(10):3232–3233. doi:10.1021/jo960471x

Li, C.Y., Wang, J.H., Luo, C.P., Ding, W.J., Cox, D.G. 2014. A new cyclopeptide with antifungal activity from the co-culture broth of two marine mangrove fungi. *Natural Product Research* 28(9). doi:10.1080/14786419.2014.887074

Li, D.L., Li, X.M., Wang, B.G. 2009. Natural anthraquinone derivatives from a marine mangrove plant-derived endophytic fungus *Eurotium rubrum*: Structural elucidation and DPPH radical scavenging activity. *Journal of Microbiology and Biotechnology* 19(7):675–680. doi:10.4014/jmb.0805.342

Li, H.Y., Wei, D.Q., Shen, M., Zhou, Z.P. 2012. Endophytes and their role in phytoremediation. *Fungal Diversity* 54:11–18.

Li, J.Y., Harper, J.K., Grant, D.M., Tombe, B.O., et al. 2001. Ambuic acid, a highly functionalized cyclohexane with antifungal activity from *Pestalotiopsis* spp. and *Monochaetia sp*. *Phytochemistry*, 56(5):463–468. https://doi.org/10.1016/S0031-9422(00)00408-8

Li, J.Y., Strobel, G.A. 2001. Jesterone and hydroxy-jesterone antioomycete cyclohexenone epoxides from the endophytic fungus *Pestalotiopsis jesteri*. *Phytochemistry* 57:261–265.

Li, J.Y., Strobel, G.A., Harper, J. et al. 2000. Cryptocin, a potent tetramic acid antimycotic from the endophytic fungus *Cryptosporiopsis* cf. *quercina*. *Organic Letters* 2:767–770.

Lim, C.S., Kim, J.Y., Choi, J.N. et al. 2010. Identification, fermentation, and bioactivity against *Xanthomonas oryzae* antimicrobial metabolites isolated from *Phomopsis longicolla* S1B4. *Journal of Microiology and Biotechnology* 20(3):494–500.

Lin, X., Huang, Y.J., Fang, M.J., Wang, J.F., Zheng, Z.H., Su, W.J. 2005. Cytotoxic and antimicrobial metabolites from marine lignicolous fungi Diaporthe sp. *FEMS Microbiology Letters* 251(1):53–58. doi:10.1016/j.femsle.2005.07.025

Lin, X., Lu, C.H., Shen, Y.M. 2008a. One new ten-embered lactone from *Phomopsis* sp. B27, an endophytic fungus of *Annona squamosa*. *Chinese Journal of Natural Medicines* 6(5):391–394.

Lin, Y., Lin, X., Zhao, P.J., Ma, J., Huang, Y.J., Shen, Y.M. 2009b. New polyketides from endophytic *Diaporthe* sp. XZ-07. *Helvetica Chimica Acta* 92(6). doi:10.1002/hlca.200800416

Lin, Z., Zhu, T., Fang, Y., Gu, Q., Zhu, W. 2008b. Polyketides from *Penicillium* sp. JP-1, an endophytic fungus associated with the mangrove plant *Aegiceras corniculatum*. *Phytochemistry* 69:1273–1278.

Lin, Z.J., Zhang, G.J., Zhu, T.J., Liu, R., Wei, H.J., Gu, Q.Q. 2009a. Bioactive cytochalasins from *Aspergillus flavipes*, an endophytic fungus associated with the mangrove plant *Acanthus ilicifolius*. *Helvetica Chimica Acta* 92:1538–1544. doi:10.1002/ hlca.200800455

Lodewyckx, C., Taghavi, S., Mergeay, M., Vangronsveld, J., et al. 2001. The effect of recombinant heavy metal-resistant endophytic bacteria on heavy metal uptake by their host plant. *International Journal of Phytoremediation*, 3(2):173–187. https://doi.org/10.1080/15226510108500055

Lodge, D.J., Fisher, P.J., Sutton, B.C. 1996. Endophytic fungi of *Manilkara bidentata* leaves in Puerto Rico. *Mycologia* 88:733–738.

Loganathan, P., Sunita, R., Parida, A.K., Nair, S. 1999. Isolation and characterization of two genetically distant groups of *Acetobacter diazotrophicus* from a new host plant *Eleusine coracana* L. *Journal of Applied Microbiology* 87:167–172.

Lösgen, S., Magull, J., Schulz, B., Draeger, S., Zeeck, A. 2008. Isofusidienols: Novel chromone-3-oxepines produced by the endo- phytic fungus *Chalara sp.* *European Journal of Organic Chemistry* 4:698–703.

Lösgen, S., Schlörke, O., Meindl, K., Herbst-Irmer, R., Zeeck, A. 2007. Structure and biosynthesis of chatocyclinones, new polyketides produced by and endosymbiotic fungus. *European Journal of Organic Chemistry* 2007: 2191–2196. https:// doi.org/10.1002/ejoc.200601020.

Lugtenberg, B.J., Caradus, J.R., Johnson, L.J. 2016. Fungal endophytes for sustainable crop production. *FEMS Microbiology Ecology* 92:fiw194.

Lugtenberg, B.J., Kamilova, F. 2009. Plant-growth-promoting rhizobacteria. *Annual Review of Microbiology* 63:541–556.

Ma, Y., Rajkumar, M., Luo, Y.M., Freitas, H. 2011a. Inoculation of endophytic bacteria on host and non-host plants-effects on plant growth and Ni uptake. *Journal of Hazardous Materials*, 195:230–237. https://doi.org/10.1016/j. jhazmat.2011.08.034

Ma, Y., Prasad, M.N.V., Rajkumar, M., Freitas, H. 2011b. Plant growth promoting rhizobacteria and endophytes accelerate phytoremediation of metalliferous soils. *Biotechnology Advances*, 29(2):248–258. https://doi.org/10.1016/j. biotechadv.2010.12.001

Maccheroni, Jr. W., Azevedo, J.L. 1998. Synthesis and secreton of phosphatases by endophytic isolates of *Colletotrichum musae* grown under conditions of nutritional starvation. *Journal of General and Applied Microbiology* 44:381–387.

Macías-Rubalcava, M.L., Hernández-Bautista, B.E., Oropeza, F., Duarte, G., González, M.C., Glenn, A.E., Hanlin, R.T, Anaya, A.L. 2010. Allelochemical effects of volatile compounds and organic extracts from *Muscodor yucatanensis*, a tropical endophytic fungus from *Bursera simaruba*. *Journal of Chemical Ecology* 36:1122–1131. doi:10.1007/s10886-010-9848-5

Macías-Rubalcava, M.L., Sánchez-Fernández, R.E. 2017a. Secondary metabolites of endophytic XYlaria species with potential applications in medicine and agriculture. *World Journal of Microbiology and Biotechnology* 33:15. doi:10.1007/s11274-016-2174-4

Macías-Rubalcava, M.L., Sánchez-Fernández, R.E. 2017b. Secondary metabolites of endophytic *Xylaria* species with potential applications in medicine and agriculture. *World Journal of Microbiology and Biotechnology* 33:15. doi:10.1007/s11274-016-2174-5

Madhaiyan, M., Poonquzhali, S., Sa, T. 2007. Metal tolerating methylotrophic bacteria reduces nickel and cadmium toxicity and promotes plant growth of tomato (*Lycopersicon esculentum* L.). *Chemosphere*, 69(2):220–228. https://doi.org/10.1016/j.chemosphere.2007.04.017

Madrigal, C., Melgarejo, P. 1995. Morphological effects of *Epiccocum nigrum* and its antibiotic flavipin on *Monilinia laxa*. *Canadian Journal of Botany* 73:425–431.

Madrigal, C., Tadeo, J.L., Melgarejo, P. 1991. Relationship between flavipin production by *Epicoccum nigrum* and antagonism against *Monilinia laxa*. *Mycological Research* 95:1375–1381.

Mahapatra, S., Banerjee, D. 2012. Structural elucidation and bioactivity of a novel exopolysaccharide from endophytic *Fusarium solani* SD5. *Carbohydrate Polymers* 90(1):683–689.

Mahapatra, S., Banerjee, D. 2013. Evaluation of in vitro antioxidant potency of exopolysaccharide from endophytic *Fusarium solani* SD5. *International Journal of Biological Macromolecules* 53:62–66.

Maitan, V.R. 1998. Isolamento e caracterizacao de actinomicetos endofiticos isolados de *Solanum lycocarpum* (lobeira). *Ms Thesis*. Universidadde Federal de Goias, Goiania, Goias, Brazil. p. 122.

Mapari, S.A.S., Meyer, A.S., Thrane, U. 2008. Evaluation of Epicoccum nigrum for growth morphology and production of natural colorants in liquid media and on a solid rice medium. *Biotechnology Letters* 30:2183–2190.

Mapari, S.A.S., Meyer, A.S., Thrane, U. 2009a. Photostability of natural orange-red and yellow fungal pigments in liquid food model systems. *Journal of Agriculture and Food Chemistry* 57:6253–6261.

Mapari, S.A.S., Meyer, A.S., Thrane, U., Frisvad, J.C. 2009b. Identification of potentially safe promising fungal cell factories for the production of polyketide natural food colorants using chemotaxonomic rationale. *Microbiology Cell Factory* 8:24.

Mapari, S.A.S., Nielsen, K.F., Larsen, T.O., Frisvad, J.C., Meyer, A.S., Thrane, U. 2005. Exploring fungal biodiversity for the production of water-soluble pigments as potential natural food colorants. *Current Opinion in Biotechnology* 16:231–238.

Mastretta, C., Taghavi, S., van der Lelie, D., Mengoni, A., et al. 2009. Endophytic bacteria from seeds of *Nicotiana tabacum* can reduce Cadmium phytotoxicity. *International Journal of Phytoremediation*, 11(3):251–267. https://doi.org/10.1080/15226510802432678

Matsuura, T. 1998. Ocorrência de actinomicetos endofíticos produtores de antibióti-cos isolados de folhas e raizes de feijão caupi (*Vigna unguiculata*). *MS thesis*. Universidade Federal de Pernambuco, Recife, Pernambuco, Brazil. p. 69.

May, R.M. 1991. A fondness for fungi. *Nature* 352:475–476.

Mejia, L.C., Rojas, E.I., Maynard, Z., Arnold, A.E., Van Bael, S., Samuels, G.J., Robbins, N., Herre, E.A. 2008. Endophytic fungi as biocontrol agents of *Theobroma cacao* pathogens. *Biological Control* 46:4–14.

Miao, Z., Wang, Y., Yu, X., Guo, B., Tang, K. 2009. A new endophytic taxane pro-duction fungus from *Taxus chinensis*. *Applied Biochemistry and Microbiology* 45:81. doi:10.1134/S0003683809010141

Mitchell, A.M. 2008. *Muscodor crispans*, a novel endophyte from *Ananas ananas-soides* in the Bolivian Amazon. *Fungal Diversity* 31:37–43.

Mitchell, A.M., Strobel, G.A., Moore, E. et al. 2010. Volatile antimicrobials from *Muscodor crispans*, a novel endophytic fungus. *Microbiology* 156:270–277.

Monggoot, S., Popluechai, S., Gentekaki, E., Pripdeevech, P. 2017. Fungal endo-phytes: An alternative source for production of volatile compounds from agar-wood oil of *Aguilaria subintegra*. *Microbial Ecology* 74:54–61. doi:10.1007/s00248-016-0908-4

Moore, F.P., Barac, T., Borremans, B., Oeyen, L., et al. 2006. Endophytic bacterial diver-sity in poplar trees growing on a BTEX-contaminated site: The characterization of isolates with potential to enhance phytoremediation. *Systematic and Applied Microbiology*, 29(7):539–556. https://doi.org/10.1016/j.syapm.2005.11.012

Moscovici, M. 2015. Present and future medical applications of microbial exopolysac-charides. *Frontiers in Microbiology*. doi:10.3389/fmicb.2015.01012

Mpika, J., Kébé, I.B., Issali, A.E., N'Guessan, F.K., Druzhinina, S., Komon-Zélazowska, M., Kubicek, C.P., Aké, S. 2009. Antagonist potential of *Trichoderma* indig-enous isolates for biological control of *Phytophthora palmivora* the causative agent of black pod disease on cocoa (*Theobroma cacao* L.) in Côte d'Ivoire. *African Journal of Biotechnology* 8:5280–5293.

Mulaw, T.B., Druzhinina, I.S., Kubicek, C.P., Atanasova, L. 2013. Novel endophytic *Trichoderma spp.* isolated from healthy *Coffea arabica* roots are capable of con-trolling coffee tracheomycosis. *Diversity* 5:750–766.

Murali, T.S., Suryanarayanan, T.S., Geeta, R. 2006. Endophytic *Phomopsis* spe-cies: Host range and implications for diversity estimates. *Canadian Journal of Microbiology* 52(7):673–680. doi:10.1139/w06-020

Murali, T.S., Suryanarayanan, T.S., Venkatesan, G. 2007. Fungal endophyte commu-nities in two tropical forests of southern India: Diversity and host affiliation. *Mycological Progress* 6:191–199.

Musetti, R., Vecchione, A., Stringher, L., Borselli, S., Zulini, L., Marzani, C., D'Ambrosio, M., Di, T.L., Pertot, I. 2006. Inhibition of sporulation and ultrastruc-tural alterations of grapevine downy mildew by the endophytic fungus *Alternaria alternata*. *Phytopathology* 96(7):689–698. doi:10.1094/phyto-96-0689

Naik, B.S., Shashikala, J., Krishnamurthy, Y.L. 2008. Diversity of fungal endophytes in shrubby medicinal plants of Malnad region, Western Ghats, Southern India. *Fungal Ecology* 1:89–93.

Narisawa, K.S. 2017. The dark septate endophytic fungus *Phialocephala fortinii* is a potential decomposer of soil organic compounds and a promoter of *Asparagus officinalis* growth. *Fungal Ecology* 28:1–10.

Nath, A., Chattopadhyay, A., Joshi, S. 2015. Biological activity of endophytic fungi of *Rauwolfia serpentina* Benth: An ethnomedicinal plant used in folk medicines in Northeast India. *Proceedings of the National Academy of Sicence India Section B: Biological Science* 85:233–240.

Nebel, G., Dragsted, J., Vanclay, J.K. 2001. Structure and floristic composition of flood plain forests in the Peruvian Amazon: II. The understorey of restinga forests. *Forest Ecological Managament* 150, 59–77.

Nettles, R., Watkins, J., Ricks, K. et al. 2016. Influence of pesticide seed treatments on rhizosphere fungal and bacterial communities and leaf fungal endophyte communities in maize and soybean. *Applied Soil Ecology* 102:61–69.

Nikolcheva, L.G., Barlocher, F. 2004. Taxon-specific fungal primers reveal unexpectedly high diversity during leaf decomposition in a stream. *Mycological Progress* 3:41–49.

Nikolcheva, L.G., Barlocher, F. 2005. Seasonal and substrate preferences of fungi colonizing leaves in streams: Traditional versus molecular evidence. *Environmental Microbiology* 7:270–280.

Nikolov, L.A., Tomlinson, P.B., Manickam, S., Endress, P.K., Kramer, E.M., Davis, C.C. 2014. Holoparasitic Rafflesiaceae possess the most reduced endophytes and yet give rise to the world's largest flowers. *Annals of Botany* 114:233–242. doi:10.1093/aob/mcu114

Noorhaida, S., Idris, A.S. 2009. *In vitro*, colonization and nursery evaluation of endophytic fungi as biological control of *Ganoderma boninense*, causal agent of basal stem rot of oil palm. In: *Proceeding PIPOC, KLCC*. Kuala Lumpur, 9–12 November 2009.

Nur, A., Muh, D.R. 2015. Isolation and characterization of endophytic fungi from medicinal plant, Buah makassar (Makassar fruit: *Brucea javanica*). *Journal of Chemical Pharmacy Research* 7:757–762.

Nutaratat, P., Srisuk, N., Arunrattiyakorn, P., Limtong, S. 2014. Plant growth-promoting traits of epiphytic and endophytic yeasts isolated from rice and sugar cane leaves in Thailand. *Fungal Biology* 118:683e694.

Oelmüller, R., Sherameti, I., Tripathi, S., et al. 2009. *Piriformospora indica*, a cultivable root endophyte with multiple biotechnological applications. *Symbiosis*, 49:1–17. https://doi.org/10.1007/s13199-009-0009-y

Ola, A.R.B., Debbbab, A., Kurtan, T. et al. 2014. Dihydroanthracenone metabolites from the endophytic fungus *Diaporthe melonis* isolated from *Annona squamosa*. *Tetrahedron Letters* 55(20):3147–3150. doi:10.1016/j.tetlet.2014.03.110

Olivares, F.L., Baldani, V.L.D., Reis, V.M., Baldani, J.I., Dobereiner, J. 1996. Occurrence of the endophytic diazotrophs *Herbaspirillum* spp. in roots, stems and leaves, predominantly of Gramineae. *Biology and Fertility of Soils* 21:197–200.

Ondeyka, J.G., Helms, G.L., Hensens, O.D. et al. 1997. Nodulisporic acid A, a novel and potent insecticide from a *Nodulisporium* sp. Isolation, structure determination and chemical transformations. *Journal of American Chemical Society* 119(38):8809–8816. doi:10.1021/ja971664k

Orlandelli, R.C., de Almeida, T.T., Alberto, R.N., Polonio, J.C., Azevedo, J.L., Pamphile, J.A. 2015. Antifungal and proteolytic activities of endophytic fungi isolated from *Piper hispidum* Sw. *Brazilian Journal of Microbiology* 46(2):359–366. doi:10.1590/S1517-838246220131042

Orlandelli, R.C., Santos, M.S., Polonio, J.C., de Azevedo, J.L., Pamphile, J.A. 2017. Use of agro-industrial wastes as substrates for α-amylase production by endophytic fungi isolated from *Piper hispidum* Sw. *Acta Scientiarum Technology Maringá* 39(3):255–261.

Orwa, C., Mutua, A., Kindt, R., Jamnadass, R., Simons, A. 2009. Agroforestree database: A tree reference and selection guide. Version 4. http://www.worldagroforestry.org/site

Palem, P.P.C., Kuriakose, G.C., Jayabaskaran, C. 2015. An endophytic fungus, *Talaromyces radicus*, isolated from *Catharanthus roseus*, produces vincristine and vinblastine, which induce apoptotic cell death. *PLoS ONE* 10(12):e0144476. doi:10.1371/journal.pone.0153111

Pan, F., Liu, Z.Q., Chen, Q., Xu, Y.W., Hou, K., Wu, W. 2016. Endophytic fungus strain 28 isolated from *Houttuynia cordata* possesses wide-spectrum antifungal activity. *Brazillian Journal of Microbiology* 47:480–488.

Pan, F., Su, T.J., Cai, S.M., Wu, W. 2017. Fungal endophyte-derived *Fritillaria unibracteata* var. *wabuensis*: Diversity, antioxidant capacities in vitro and relations to phenolic, flavonoid or saponin compounds. *Scientific Reports*, 7:42008. https://doi.org/10.1038/srep42008

Pandey, A.K., Reddy, M.S., Suryanarayanan, T.S. 2003. ITS-RFLP and ITS sequence analysis of a foliar endophytic *Phyllosticta* from different tropical trees. *Mycological Research* 107(4):439–444. doi:10.1017/S0953756203007494

Parshikov, I., Heinze, T., Moody, J. et al. 2001. The fungus *Pestalotiopsis guepini* as a model for biotransformation of ciprofloxacin and norfloxacin. *Applied Microbiology and Biotechnology* 56:474–477. doi:10.1007/s002530100672

Parthasarathy, R., Sathiyabama, M. 2015. Lovastatin-producing endophytic fungus isolated from a medicinal plant *Solanum xanthocarpum*. *Natural Product Research* 29:2282–2286.

Patil, M.G., Pagare, J., Patil, S.N., Sidhu, A.K. 2015. Extracellular enzymatic activities of endophytic fungi isolated from various medicinal plants. *International Journal of Current Microbiology Applied Science* 4:1035–1042.

Pedras, M.S.C., Ahiahonu, P.W.K. 2005. Metabolism and detoxification of phytoalexins and analogs by phytopathogenic fungi. *Phytochemistry*, 66(4):391–411. https://doi.org/10.1016/j.phytochem.2004.12.032

Pedraza, R.O., Bellone, C.H., de Bellone, C.S. et al. 2009. *Azospirillum* inoculation and nitrogen fertilization effect on grain yield and on the diversity of endophytic bacteria in the phyllosphere of rice rainfed crop. *European Journal of Soil Biology* 45:36–43.

Pereira, J.O., Azevedo, J.L., Petrini, O. 1993. Endophytic fungi of stylosanthes: A first report. *Mycologia* 85:362–364.

Pereira, J.O., Carneiro-Vieira, M.L., Azevedo, J.L. 1999. Endophytic fungi from *Musa acuminata* and their reintroduction into axenic plants. *World Journal of Microbiology and Biotechnology* 15:37–40.

Petrini, O. 1986. Taxonomy of endophytic fungi in aerial plant tissues. In: Fokkema, N.J., van den Heuvel, J. (eds). *Microbiology of the Phyllosphere*. Cambridge University Press, Cambridge, UK., pp. 175–187.

Petrini, O. 1991. Fungal endophytes in tree leaves. In: Andrews, J.H., Hirano, S.S. (eds). *Microbial Ecology of Leaves*. Springer, New York, pp. 179–197.

Petrini, O., Dreifuss, M. 1981. Endophytische pilze in epiphytischen Aracea, Bromeliaceae und Orchidaceae. *Sydowia* 34:135–148.

Pinto, L.S.R.C., Azevedo, J.L., Pereira, J.O., Vieira, M.L.C., Labate, C.A. 2000. Symptomless infection of banana and maize by endophytic fungi impairs photosynthetic efficiency. *New Phytologist* 147:609–615.

Pliego, C., Kamilova, F., Lugtenberg, B. 2011. Plant growth-promoting bacteria: Fundamentals and exploitation. In: Maheshwari, D.K. (ed). *Bacteria in Agrobiology: Crop Ecosystems.* Springer, Berlin Heidelberg, pp. 295–343.

Prachya, S., Wiyakrutta, S., Sriubolmas, N. et al. 2007. Cytotoxic mycoepoxydiene derivativesfrom an endophytic fungus *Phomopsis sp.* isolated from *Hydnocarpus anthelminthicus. Planta Medica* 73(13):1418–1420. doi:10.1055/s-2007-990240

Pu, X., Qu, X., Chen, F., Bao, J., Zhang, G., Luo, Y. 2013. Camptothecin-producing endophytic fungus *Trichoderma atroviride* LY357: Isolation, identification, and fermentation conditions optimization for camptothecin production. *Applied Microbiology and Biotechnology* 97(21):9365–9375. doi:10.1007/s00253-013-5163-8

Puente, M.E., Li, C.Y., Bashan, Y. 2009. Endophytic bacteria in cacti seeds can improve the development of cactus seedlings. *Environmental and Experimental Botany* 66(3):402–408. doi:10.1016/j.envexpbot.2009.04.007

Qian, C.D., Fu, Y.H., Jiang, F.S., Xu, Z.H., Cheng, D.Q., Ding, B., Gao, C.X., Ding, Z.S. 2014. *Lasiodiplodia* sp. ME4-2, an endophytic fungus from the floral parts of *Viscum coloratum,* produces indole-3-carboxylic acid and other aromatic metabolites. *BMC Microbiology* 14:297. http://www.biomedcentral.com/1471-2180/14/297

Qin, S., Chen, H.H., Zhao, G.Z., Li, J., Zhu, W.Y., Xu, L.H., Jiang, J.H., Li, W.J. 2012. Abundant and diverse endophytic actinobacteria associated with medicinal plant *Maytenus austroyunnanensis* in Xishuangbanna tropical rainforest revealed by culture-dependent and culture-independent methods. *Environmental Microbiology Reports* 4(5):522–531. doi:10.1111/j.1758-2229.2012.00357.x

Qin, S., Li, J., Chen, H.H., Zhao, G.Z., Zhu, W.Y., Jiang, C.L., Xu, L.H., Li, W.J. 2009. Isolation, diversity, and antimicrobial activity of are actinobacteria from medicinal plants of tropical rain forests in Xishuangbanna, China. *Applied and Environmental Microbiology* 75(19):6176–6186. doi:10.1128/AEM.01034-09

Qiu, M., Xie, R.S., Shi, Y., Chen, H.M., Wen, Y.L., Gao, Y.S., Hu, X.F. 2010. Isolation and identification of endophytic fungus SX01, a red pigment producer from *Ginkgo biloba* L. *World Journal of Microbiology and Biotechnology* 26:993–998. doi:10.1007/s11274-009-0261-6

Raheman, F., Deshmukh, S., Ingle, A., Gade, A., Rai, M. 2011. Silver nanoparticles: Novel antimicrobial agent synthesized from an endophytic fungus *Pestalotia* sp. isolated from leaves of *Syzygium cumini* (L). *Nanotechnology Biomedical Engineering*, 3(3):174–178. doi:10.5101/nbe.v3i3.p174-178

Rajkumar, M., Prasad, M.N.V., Freitas, H. 2010. Potential of siderophore-producing bacteria for improving heavy metal phytoextraction. *Trends in Biotechnology* 28:142–149.

Rajulu, M.B.G., Thirunavukkarasu, N., Babu, A.G., Aggarwal, A., Suryanarayanan, T.S., Reddy, M.S. 2013. Endophytic xylariaceae from the forests of Western Ghats, Southern India: Distribution and biological activities. *Mycology* 4:29–37. doi:10.1080/21501203.2013.776648

Rajulu, M.B.G., Thirunavukkarasu, N., Suryanarayanan, T.S. 2011. Chitinolytic enzymes from endophytic fungi. *Fungal Diversity* 47(1):43–53.

Ramírez-Coronel, M.A., Viniegra-Gonzalez, G., Darvill, A., Augur, C. 2003. A novel tannase from *Aspergillus niger* with β-glucosidase activity. Microbiology, 149(10):2941–2946.

Rao, M.B., Tanksale, A.M., Ghatge, M.S., Deshpande, V.V. 1998. Molecular and biotechnological aspects of microbial proteases. *Microbiology and Molecular Biology Reviews*, 62(3):597–635.

Redman, R.S., Sheehan, K.B., Stout, R.G. et al. 2002. Thermotolerance generated by plant/fungal symbiosis. *Science* 298:1581.

Reis, V.M., Olivares, F.L., Dobereiner, J. 1994. Improved methodology for isolation of *Acetobacter diazotrophicus* and confirmation of its endophytic habitat. *World Journal of Microbiology and Biotechnology* 10:401–405.

Reino, J.L., Guerrero, R.F., Hernández-Galán, R., et al. 2008. Secondary metabolites from species of the biocontrol agent *Trichoderma*. *Phytochemistry Reviews* 7:89–123. https://doi.org/10.1007/s11101-006-9032-2

Rhoden, S.A., Garcia, A., Rubin-Filho, C.J., Azevedo, J.L., Pamphile, J.A. 2012. Phylogenetic diversity of endophytic leaf fungus isolates from the medicinal tree *Trichilia elegans* (Meliaceae). *Genetics and Molecular Research*, 11:2513–2522.

Ribeiro, S.F.L., daCosta Garcia, A., dos Santos, H.E.D., Montoya, Q.V., Rodirgues, A., deOliveira, J.M., deOliveira, C.M. 2018. Antimicrobial activity of crude extracts of endophytic fungi from *Oryctanthus alveolatus* (Kunth) Kuijt (Mistletoe). *African Journal of Microbiology Research*, 12(11):263–268.

Rodrigues, K.F. 1994. The foliar fungal endophytes of the Amazonian palm *Euterpe oleracea*. *Mycologia* 86:376–385.

Rodrigues, K.F., Leuchtmann, A., Petrini, O. 1993. Endophytes species of xylaria: Cultural and isozymic studies. *Sidowia* 45:116–138.

Rodrigues, K.F., Samuels, G.J. 1990. Preliminary study of endophytic fungi in a tropical palm. *Mycological Research* 94:827–830.

Rodrigues, K.F., Samuels, G.J. 1992. *Idriella* species endophytic in palms. *Mycotaxon* 43:271–276.

Rodrigues, K.F., Samuels, G.J. 1999. Fungal endophytes of *Spondias mombin* leaves in Brazil. *Journal of Basic Microbiology* 39:131–135.

Rodriguez, R.J., Henson, J., Van Volkenburgh, E. et al. 2008. Stress tolerance in plants via habitat-adapted symbiosis. *ISME Journal* 2:404–416.

Ronsberg, D., Debbbab, A., Mandi, A., Vasylyeva, V., et al. 2013. Pro-apoptotic and immunostimulatory tetrahydroxanthone dimers from the endophytic fungus *Phomopsis longicolla*. *Journal of Organic Chemistry*, 78(24):12409–12425. https://doi.org/10.1021/jo402066b

Rosa, L.H., Gonçalves, V.N., Caligiorne, R.B. et al. 2010. Leishmanicidal, trypanocidal, and cytotoxic activities of endophytic fungi associated with bioactive plants in Brazil. *Brazilian Journal of Microbiology* 41(2):420–430. doi:10.1590/S1517-83822010000200024

Rosa, L.H., Tabanca, N., Techen, N. et al. 2012. Antifungal activity of extracts from endophytic fungi associated with *Smallanthus* maintained in vitro as autotrophic cultures and as pot plants in the greenhouse. *Canadian Journal of Microbiology* 58(10):1202–1211. doi:10.1139/w2012-088

Rungjindamai, N., Pinruan, U., Choeyklin, R., Hattori, T., Jones, E.B.G. 2008. Molecular characterization of basidiomycetous endophytes isolated from leaves, rachis and petioles of the oil palm, *Elaeis guineensis*, in Thailand. *Fungal Diversity* 33:139–161.

Rukachaisirikul, V., Sommart, U., Phongpaichit, S., Sakayaroj, J., Kirtikara, K. 2008. Metabolites from the endophytic fungus *Phomopsis* sp. PSU-D15. *Phytochemistry*, 69(3):783–787. doi:10.1016/j.phytochem.2007.09.006

Saleem, M., Arshad, M., Hussain, S., Bhatti, A.S. 2007. Perspective of plant growth promoting rhizobacteria (PGPR) containing ACC deaminase in stress agriculture. *Journal of Industrial Microbiology & Biotechnology* 34:635–648.

Samuels, G.J., Dodd, S.L., Lu, B.S., Petrini, O., Schroers, H.J., Druzhinina, I.S. 2006a. The *Trichoderma koningii* aggregate species. *Studies in Mycology*, 56:67–133. https://doi.org/10.3114/sim.2006.56.03

Samuels, G.J., Ismaiel, A. 2009. *Trichoderma evansii* and *T. lieckfeldtiae*: two new *T. hamatum*-like species. *Mycologia*, 101(1):142–156. https://doi.org/10.3852/08-161

Samuels, G.J., Ismaiel, A., Mulaw, T.B., Szakacs, G., Druzhinina, I.S., Kubicek, C.P., Jaklitsch, W.M. 2012. The *Longibrachiatum* clade of *Trichoderma*: A revision with new species. *Fungal Diversity* 55:77–108.

Samuels, G.J., Pardo-Schultheiss, R., Hebbar, K.P., Lumsden, R.D., et al. 2000. *Trichoderma stromaticum* sp. nov., a parasite of the cacao witches broom pathogen. *Mycological Research*, 104(6):760–764. https://doi.org/10.1017/S0953756299001938

Samuels, G.J., Suarez, C., Solis, K., Holmes, K.A., Thomas, S.E., Ismaiel, A., Evans, H.C. 2006b. *Trichoderma theobromicola* and *T. paucisporum*: Two new species isolated from cacao in South America. *Mycological Research*, 110(4):381–392. https://doi.org/10.1016/j.mycres.2006.01.009

Sanchez-Ortiz, B.L., Sanchez-Fernandez, R.E., Duarte, G., Lappe-Oliveras, P., Macias-Rubalcava, M.L. 2016. Antifungal, anti-oomycete and phytotoxic effects of volatile organic compounds from the endophytic fungus *Xylaria* sp. strain PB3f3 isolated from *Haematoxylon brasiletto*. *Journal of Applied Microbiology*, 120(5):1313–1325. https://doi.org/10.1111/jam.13101

Santosfo, F., Fill, T.P., Nakamura, J., Monteiro, M.R., Rodriguesfo, E. 2011. Endophytic fungi as a source of biofuel precursors. *Journal of Microbiology and Biotechnology* 21(7):728–733. doi:10.4014/jmb.1010.10052

Sarang, H., Rajani, P., Vasanthakumari, M.M., Kumara, P.M., Siva, R., Ravikanth, G., Shaanker, R.U. 2017. An endophytic fungus, *Gibberella moniliformis* from *Lawsonia inermis* L. produces lawsone, an orange-red pigment. *Antonie van Leeuwenhoek* 110:853–862. doi:10.1007/s10482-017-0858-y

Saravanan, V.S., Madhaiyan, M., Thangaraju, M. 2007. Solubilization of zinc compounds by the diazotrophic, plant growth promoting bacterium *Gluconacetobacter diazotrophicus*. *Chemosphere*, 66(9):1794–1798. https://doi.org/10.1016/j.chemosphere.2006.07.067

Sarquis, M.I.M., Oliveira, E.M.M., Santos, A.S., daCosta, G.L. 2004. Production of L-asparaginase by filamentous fungi. *Memorias do Instituto Oswaldo Cruz* 99(5). doi:10.1590/S0074-02762004000500005

Schmelz, E.A., Engelberth, J., Alborn, H.T., O'Donnell, P., Sammons, M., Toshima, H., Tumlinson, J.H. III. 2003. Simultaneous analysis of phytohormones, phytotoxins, and volatile organic compounds in plants. *Proceedings of the National Academy of Sciences of the United States of America* 100:10552–10557.

Schulz, B.J.E. 2006. Mutualistic interactions with fungal root endophytes. In: Schulz B.J.E., Boyle, C.J.C., Sieber, T.N. (eds). *Microbial Root Endophytes*, Vol. 9. Springer, Berlin, Heidelberg, pp. 261–279.

Schulz, B., Boyle, C. 2005. The endophytic continuum. *Mycological Research* 109(6):661–686.

Schulz, B., Boyle, C. 2006. What are endophytes? In: Schulz, B.J.E., Boyle, C.J.C., Sieber, T.N. (eds). *Microbial Root Endophytes.* Springer, Berlin, Germany. pp. 1–13.

Schulz, B., Boyle, C., Draeger, S., Rommert, A.K., Krohn, K. 2002. Endophytic fungi: A source of biologically active secondary metabolites. *Mycological Research* 106:996–1004.

Sebastianes, F.L.S., Romão-Dumaresq, A.S., Lacava, P.T., Harakava, R., Azevedo, J.L., DeMelo, I.S., Pizzirani-Kleiner, A.A. 2013. Species diversity of culturable endophytic fungi from Brazilian mangrove forests. *Current Genetics*, 59:153–166. https://doi.org/10.1007/s00294-013-0396-8

Seena, S., Wynberg, N., Barlocher, F. 2008. Fungal diversity during leaf decomposition in a stream assessed through clone libraries. *Fungal Diversity* 30:1–14.

Sharma, K.K., Saikia, R., Kotoky, J., Kalita, J.C., Devi, R. 2011. Antifungal activity of *Solanum melongena* L. *Lawsonia inermis* L. and *Justicia gendarussa* B. against Dermatophytes. *International Journal of PharmTech Research*, 3:1635–1640.

Sheng, X.F., Xia, J.J., Jiang, C.Y., He, L.Y., Qian, M. 2008a. Characterization of heavy metal-resistant endophytic bacteria from rape (*Brassica napus*) roots and their potential in promoting the growth and lead accumulation of rape. *Environmental Pollution*, 156(3):1164–1170. https://doi.org/10.1016/j.envpol.2008.04.007

Sheng, X.F., Jiang, C.Y., He, L.Y. 2008b. Characterization of plant growth-promoting *Bacillus edaphicus* NBT and its effect on lead uptake by India mustard in a lead-amended soil. *Canadian Journal of Microbiology*, 54(5):417–422. https://doi.org/10.1139/W08-020

Shi, J.L., Liu, C., Liu, L., Yang, B., Zhang, Y. 2012. Structure identification and fermentation characteristics of pinoresinol diglucoside produced by *Phomopsis* sp. isolated from *Eucommia ulmoides* Oliv. *Applied Microbiology and Biotechnology*, 93:1475–1483. https://doi.org/10.1007/s00253-011-3613-8

Shibuya, H., Agusta, A., Ohashi, K., Maehara, S., Simanjuntak, P. 2005. Biooxidation of (+)-catechin and (–)-epicatechin into 3,4-dihydroxyflavan derivatives by the endophytic fungus *Diaporthe* sp. Isolated from a tea plant. *Chemical and Pharmaceutical Bulletin* 53:866–867.

Shrivastava, S., Varma, A. 2014. From *Piriformospora indica* to rootonic: A review. *African Journal of Microbiology Research* 8:2984–2992.

Shweta, S., Shaanker, R.U. 2010. Endophytic fungal strains of *Fusarium solani*, from *Apodytes dimidiate* E. Mey. Ex Arn (Icacinaceae) produce camptothecin, 10-hydroxycamptothecin and 9-methoxycamptothecin. *Phytochemistry* 71(1):117–122.

Siciliano, S.D., Fortin, N., Mihoc, A., Wisse, G., Labelle, S., et al. 2001. Selection of specific endophytic bacterial genotypes by plants in response to soil contamination. *Applied and Environmental Microbiology*, 67:2469–2475. doi:10.1128/AEM.67.6.2469-2475.2001

Silva, L.F., Freire, K.T.L.S., Araújo-Magalhães, G.R., Agamez-Montalvo, G.S., Sousa, M.A., Costa-Silva, T.A., Paiva, L.M., Pessoa-Junior, A., Bezerra, J.D.P., Souza-Motta, C.M. 2018. *Penicillium* and *Talaromyces* endophytes from *Tillandsia catimbauensis*, a bromeliad endemic in the Brazilian tropical dry forest, and their potential for l-asparaginase production. *World Journal of Microbiology and Biotechnology* 34:162. doi:10.1007/s11274-018-2547-z

Sim, C.S.F., Chen, S.H., Ting, A.S.Y. 2019. Endophytes: Emerging tools for the bioremediation of pollutants. In: Bharagava, R., Chowdhary, P. (eds). *Emerging and Eco-Friendly Approaches for Waste Management*. Springer, Singapore. pp. 189–217. Print ISBN: 978-981-10-8668-7, Online ISBN: 978-981-10-8669-4. doi:10.1007/978-981-10-8669-4_10

Sim, C.S.F., Cheow, Y.L., Ng, S.L., Ting, A.S.Y. 2016. Endophytes from *Phragmites* for metal removal: Evaluating their metal tolerance, adaptive tolerance behaviour and biosorption efficacy. *Desalination and Water Treatment* 57(15):6959–6966. doi:10.1080/19443994.2015.1013507

Sim, C.S.F., Cheow, Y.L., Ng, S.L., Ting, A.S.Y. 2018. Discovering metal-tolerant endophytic fungi from the phytoremediator plant *Phragmites*. *Water Air Soil Pollution* 229:68. doi:10.1007/s11270-018-3733-1

Sim, J.H., Khoo, C.H., Lee, L.H., Cheah, Y.K. 2010. Molecular diversity of fungal endophytes isolated from *Garcinia mangostana* and *Garcinia parvifolia*. *Journal of Microbiology and Biotechnology* 20(4):651–658. doi:10.4014/jmb.0909.09030

Smith, S.A., Tank, D.C., Boulanger, L.A. et al. 2008. Bioactive endophytes warrant intensified exploration and conservation. *PLoS ONE* 3(8):e3052. doi:10.1371/journal.pone.0003052

Soleimani, M., Hajabbasi, M.A., Afyuni, M, et al. 2010. Effect of endophytic fungi on cadmium tolerance and bioaccumulation by *Festuca arundinacea* and *Festuca pratensis*. *International Journal of Phytoremediation* 12(6):535–549. doi:10.1080/15226510903353187

Soliman, S.S.M., Raizada, M.N. 2013. Interactions between co-inhabiting fungi elicit synthesis of Taxol from an endophytic fungus in host *Taxus* plants. *Frontiers in Microbiology*, 4:3. https://doi.org/10.3389/fmicb.2013.00003

Soliman, S.S.M., Raizada, M.N. 2018. Darkness: A crucial factor in fungal taxol production. *Frontiers in Microbiology*, 9:353. https://doi.org/10.3389/fmicb.2018.00353

Song, F., Wu, S.H., Zhai, Y.Z., Xuan, Q.C., Wang, T. 2014. Secondary metabolites from the genus *Xylaria* and their bioactivities. *Chemistry and Biodiversity*, 11(5):673–694. https://doi.org/10.1002/cbdv.201200286

Sopalun, K., Strobel., G.A., Hess, W.M., Worapong, J. 2003. A record of *Muscodor albus*, an endophyte from *Myristica fragrans* in Thailand. *Mycotaxon*, 88:239–248.

Southcott, K.A., Johnson, J.A. 1997. Isolation of endophytes from two species of palm from Bermuda. *Canadian Journal of Microbiology* 43:789–792.

Spagnoletti, F.N., Lavado, R.S. 2017. Dark septate endophytes present different potential to solubilize calcium, iron and aluminium phosphates. *Applied Soil Ecology* 111:25–32.

Stadler, M., Keller, N.P. 2008. Paradigm shifts in fungal secondary metabolite research. *Mycological Research* 112(2):127–130.

Stadler, M., Schulz, B. 2009. High energy biofuel from endophytic fungi? *Trends in Plant Science* 14(7):353–355.

Statzell-Tallman, A., Belloch, C., Fell, J.W. 2008. *Kwoniella mangroviensis* gen. nov., sp.nov. (Tremellales, Basidiomycota), a teleomorphic yeast from mangrove habitats in the Florida Everglades and Bahamas. *FEMS Yeast Research* 8(1):103–113. doi:10.1111/j.1567-1364.2007.00314.x

Stierle, A.A., Stierle, D.B. 2000. Bioactive compounds from four endophytic *Peniciullium* sp. of a northwest Pacific yew tree. *Studies in Natural Products Chemistry* 24(e):933–977.

Stierle, A., Strobel, G., Stierle, D. 1993. Taxol and taxane production by *Taxomyces andreanae*, an endophytic fungus of Pacific yew. *Science* 260(5105):214–216.

Stinson, M., Ezra, D., Hess, W.M. et al. 2003. An endophytic *Gliocladium* sp. of *Eucryphia cordifolia* producing selective volatile antimicrobial compounds. *Plant Science* 165:913–922.

Stone, J.K., Bacon, C.W., White, J.F. 2000. An overview of endophytic microbes: Endophytism defined. In: Bacon, C.W., White, J.F. (eds). *Microbial Endophytes*. C. W. Bacon Dekker, New York, pp. 3–30.

Strobel, G.A. 2002. Rainforest endophytes and bioactive products. *Critical Review of Biotechnology* 22(4):315–333. doi:10.1080/07388550290789531

Strobel, G.A. 2006. Harnessing endophytes for industrial microbiology. *Current Opinion in Microbiology* 9:240–244.

Strobel, G.A., Daisy, B. 2003. Bioprospecting for microbial endophytes and their natural products. *Microbiology Molecular Biology Reviews* 67:491–502.

Strobel, G.A., Daisy, B., Castillo, U., Harper, J. 2004. Natural products from endophyticmicroorganisms. *Journal of Natural Products* 67:257–268.

Strobel, G.A., Dirkse, E., Sears, J. et al. 2001. Volatile antimicrobials from *Muscodor albus*, a novel endophytic fungus. *Microbiology* 147:2943–2950.

Strobel, G.A., Ford, E., Worapong, J. et al. 2002. Isopestacin, an isoben-zofuranone from *Pestalotiopsis microspora*, possessing anti-fungal and antioxidant activities. *Phytochemistry* 60:179–183.

Strobel, G.A., Knighton, B., Kluck, K., Ren, Y., Livinghouse, T., Griffin, M., Spakowicz, D., Sears, J. 2008. The production of mycodiesel hydrocarbons and their derivatives by the endophytic fungus *Gliocladium roseum* (NRRL 50072). *Microbiology* 154(11):3319–3328. doi:10.1099/mic.0.30824-0

Strobel, S.A., Strobel, G.A. 2007. Plant endophytes as a platform for discovery-based undergraduate science education. *Natural Chemistry Biology* 3:356–359.

Sun, C., Johnson, J.M., Cai, D., Sherameti, I., Oelmüller, R., Lou, B. 2010. *Piriformospora indica* confers drought tolerance in Chinese cabbage leaves by stimulating antioxidant enzymes, the expression of drought-related genes and the plastid-localized CAS protein. *Journal of Plant Physiology* 167(12):1009–1017. doi:10.1016/j.jplph.2010. 02.013

Sun, R.Y., Liu, Z.C., Fu, K.H., Fan, L.L., Chen, J. 2012. *Trichoderma* biodiversity in China. *Journal of Applied Genetics*, 53:343–354. https://doi.org/10.1007/s13353-012-0093-1

Sundin, G.W., Murillo, J. 1999. Functional analysis of the *Pseudomonas syringae* rulAB determinant in tolerance to ultraviolet B (290–320 nm) radiation and distribution of rulAB among *P. syringae* pathovars. *Environmental Microbiology* 1:75–87.

Suresh, H.S., Bhat, H.R., Dattaraja, H.S., Sukumar, R. 1999. Flora of mudumalai wildlife sanctuary, Nilgiris, Tamil Nadu. *Technical Representative* 64, Center for Ecological Sciences, Indian Institute of Science, Bangalore, India.

Suryanarayanan, T.S. 2013. Endophyte research: Going beyond isolation and metabolite documentation. *Fungal Ecology* 6:561–568.

Suryanarayanan, T.S., Kumaresan, V., Johnson, J.A. 1998. Foliar fungal endophytes from two species of the mangrove *Rhizophora*. *Canadian Journal of Microbiology* 44:1003–1006.

Suryanarayanan, T.S., Murali, T.S. 2006. Incidence of *Leptosphaerulina crassiasca* in symptomless leaves of peanut in southern India. *Journal of Basic Microbiology* 46:305–309.

Suryanarayanan, T.S., Murali, T.S., Thirunavukkarasu, N., Rajulu, M.B.G., Venkatesan, G., Sukumar, R. 2011. Endophytic fungal communities in woody perennials of three tropical forest types of the Western Ghats, Southern India. *Biodiversity and Conservation* 20:913–928.

Suryanarayanan, T.S., Murali, T.S., Venkatesan, G. 2002. Occurrence and distribution of fungal endophytes in tropical forests across a rainfall gradient. *Canadian Journal of Botany* 80:818–826.

Suryanarayanan, T.S., Thirunavukkarasu, N., Rajulu, M.B.G., Gopalan, V. 2012. Fungal endophytes: An untapped source of biocatalysts. *Fungal Diversity* 54:19–30. doi:10.1007/s13225-012-0168-7

Suryanarayanan, T.S., Thirunavukkarasu, N., Rajulu, M.B.G., Sasse, F., Jansen, R., Murali, T.S. 2009. Fungal endophytes and bioprospecting. *Fungal Biology Reviews* 23:9–19.

Tan, R., Zou, W. 2001. Endophytes: A rich source of functional metabolites. *Natural Product Report* 18:448–459.

Tang, Y.J., Zhao, W., Li, H.M. 2011. Novel tandem biotransformation process for the biosynthesis of a novel compound, 4-(2,3,5,6-Tetramethylpyrazine-1)-4′-Demethylepipodophyllotoxin. *Applied and Environmental Microbiology* 77(9):3023–3034. doi:10.1128/AEM.03047-10

Tao, G., Liu, Z.Y., Hyde, K.D., Lui, X.Z., Yu, Z.N. 2008. Whole rDNA analysis reveals novel and endophytic fungi in *Bletilla ochracea* (Orchidaceae). *Fungal Diversity* 33:101–122.

Taylor, J.E., Hyde, K.D., Jones, E.B.G. 1999. Endophytic fungi associated with the temperate palm *Trachycarpus fortunei* within and outside its natural geographic range. *New Phytologist* 142:335–346.

Thirunavukkarasu, N., Suryanarayanan, T.S., Murali, T.S., Ravishankar, J.P., Gummadi, S.N. 2011. L-asparaginase from marine derived fungal endophytes of seaweeds. *Mycosphere* 2:147–155.

Ting, A.S.Y. 2014. Biosourcing endophytes as biocontrol agents of wilt diseases. In: Verma, V., Gange, A. (eds). *Advances in Endophytic Research*. Springer, New Delhi Biosourcing Endophytes as Biocontrol Agents of Wilt Diseases. doi:10.1007/978-81-322-1575-2_15

Ting, A.S.Y., Lee, M.V.J., Chow, Y.Y., Cheong, S.L. 2016. Novel exploration of endophytic *Diaporthe* sp. for the biosorption and biodegradation of triphenylmethane dyes. *Water, Air, Soil Pollution* 227:109. doi:10.1007/s11270-016-2810-6

Toghueo, R.M.K., Zabalgogeazcoa, I., Vázquez de Aldana, B.R., Boyom, F.F. 2017. Enzymatic activity of endophytic fungi from the medicinal plants *Terminalia catappa*, *Terminalia mantaly* and *Cananga odorata*. *South African Journal of Botany* 109:146–153.

Unterseher, M., Gazis, R., Chaverri, P., Guarniz, C.F.G., Tenorio, D.H.Z. 2013. Endophytic fungi from Peruvian highland and lowland habitats form distinctive and host plant-specific assemblages. *Biodiversity and Conservation* 22:999–1016.

U'Ren, J.M., Lutzoni, F., Miadlikowska, J., Arnold, A.E. 2010. Community analysis reveals close affinities between endophytic and endolichenic fungi in mosses and lichens. *Microbial Ecology* 60:340–353.

U'Ren, J.M., Lutzoni, F., Miadlikowska, J., Laetsch, A., Arnold, A.E. 2012. Host-and geographic structure of endophytic and endolichenic fungi at a continental scale. *American Journal of Botany* 99:898–914.

Van Bael, S.A., Valencia, M., Rojas, E., Gomez, N., Windsor, D.M., Herre, E.A., 2009. Effects of foliar endophytic fungi on the preference and performance of a leaf beetle, *Chelymorpha alternans* Boheman (Chrysomelidae: Cassidinae). *Biotropica* 41:221–225.

Varughese, T., Rios, N., Higginbotham, S. et al. 2012. Antifungal depsidone metabolites from *Cordyceps dipterigena*, an endophytic fungus antagonistic to the phytopathogen *Gibberella fujikuroi. Tetrahedron Letters* 53:1624–1626.

Vaz, A.B.M., Mota, R.C., Bomfim, M.R.Q., Vieira, M.L.A., Zani, C.L., Rosa, C.A., Rosa, L.H. 2009. Antimicrobial activity of endophytic fungi associated with Orchidaceae in Brazil. *Canadian Journal of Microbiology* 55(12):1381–1391. doi:10.1139/W09-101

Vega, F.E., Posada, F., Aime, M.C., Pava-Ripoll, M., Infante, F., Rehner, S.A. 2008. Entomopathogenic fungal endophytes. *Biological Control* 46:72–82.

Venekar, S.A., Mishra, B.D., Sreekumar, E.S., Deshmukh, S.K., Fiebig, H., Kelter, G., Maier, A. 2014. Anticancer activity of new depsipeptide compound isolated from an endophytic fungus. *Journal of Antibiotic* 67:697–701.

Vieira, M.L.A., Hughes, A.F.S., Gil, V.B., Vaz, A.B.M., Alves, T.M.A., Zani, C.L., Rosa, C.A., Rosa, L.H. 2012. Diversity and antimicrobial activities of the fungal endophyte community associated with the traditional Brazillian medicinal plant *Solanum cernuum* Vell. (Solanacea). *Canadian Journal of Microbiology* 58:54–66.

Vieira, M.L.A., Johann, S., Hughes, F.M., Rosa, C.A., Rosa, L.H. 2014. The diversity and antimicrobial activity of endophytic fungi associated with medicinal plant *Baccharis trimera* (Asteraceae) from the Brazillian savannah. *Canadian Journal of Microbiology*, 60(12):847–856. https://doi.org/10.1139/cjm-2014-0449

Wagenaar, M.M., Clardy, J. 2001. Dicerandrols, new antibiotic and cytotoxic dimers produced by the fungus *Phomopsis longicolla* isolated from an endangered mint. *Journal of Natural Product* 64(8):1006–1009. doi:10.1021/np010020u

Wang, M., Zhang, W., Xu, W. et al. 2016. Optimization of genome shuffling for high-yield production of the antitumour deactylmycoepoxydiene in an endophytic fungus of mangrove plants. *Applied Microbiology and Biotechnology* 100(17):7491–7498. doi:10.1007/s00253-016-7457-0

Wang, X.J., Min, C.L., Ge, M. et al. 2014. An endophytic sanguinarine-producing fungus from Macleaya cordata, Fusarium proliferatum BLH51. *Current Microbiology* 68(3):336–341. doi:10.1007/s00284-013-0482-7

Wang, Y., Li, H., Feng, G., Du, L., Zeng, D. 2017. Biodegradation of diuron by an endophytic fungus *Neurospora intermedia* DP8-1 isolated from sugarcane and its potential for remediating diuron-contaminated soils. *PLoS ONE* 12(8):e0182556. doi:10.1371/journal.pone.0182556

Wangun, H.V.K., Hertweck, C. 2007. Epicoccarines A, B and epipyridone: Tetramic acids and pyridone alkaloids from an *Epicoccum* sp. Associated with the tree fungus *Pholiota squarrosa. Organic and Biomolecular Chemistry* 5:1702–1705.

Wani, M.C., Taylor, H.L., Wall, M.E., Coggon, P., McPhail, A.T. 1971. Plant antitumour agents. VI. Isolation and structure of taxol, a novel antileukemic and antitumour agent from *Taxus brevifolia*. *Journal of American Chemical Society* 93(9):2325–2327.

Webster, J., Weber, R. 2007. *Introduction to Fungi* (3rd edn). Cambridge University Press, New York.

Wei, S., We, S.Z., Wang, T.T. et al. 2015. *Sphingomonas hengshuiensis* sp. nov., isolated from lake wetland. *International Journal of Systematic Evolutionary MIcrobiology* 65:4644–4649. doi:10.1099/ijsem.0.000626

Wei, Y., Yu, H.H., Guan, X.S. et al. 2014. Genetic diversity of endophytic bacteria of the manganese-hyperaccumulating plant Phytolacca americana growing at a manganese mine. *European Journal of Soil Biology* 62:15–21. doi:10.1016/j.ejsobi.2014.02.011

Wei, Y.M., Liu, L., Zhou, X.W. et al. 2012. Engineering taxol biosynthetic pathway for improving taxol yield in taxol-producing endophytic fungus EFY21 *(Ozonium sp.)*. *African Journal of Biotechnology* 11(37):9094–9101. doi:10.5897/AJB10.1896

Weyens, N., Thijs, S., Popek, R., Witters, N., Przybysz, A., Espenshade, J., Gawronska, H., Vangromsveld, J., Gawronski, S.W. 2015. The role of plant-microbe interactions and their exploitation for phtoremediation of air pollutants. *International Journal of Molecular Science* 16:25576–25604. doi:10.3390/ijms161025576

Whalley, A.J.S., Edwards, R.L. 1995. Secondary metabolites and systematic arrangement within the Xylariaceae. *Canadian Journal of Botany* 73:S802–S810.

Widowati, T., Bustanussalam, B., Sukiman, H., Simanjuntak, P. 2016. Isolasi dan identifikasi kapang endofit dari Tanaman Kunyit (*Curcuma longa* L.) sebagai penghasil antioksidan. *Biopropal Industri* 7:9–16.

Wilson, D. 2000. Ecology of woody plant endophytes. In: Bacon, C.W., White, Jr. J.F. (eds). *Microbial Endophytes*. Marcel Dekker, Inc., New York, pp. 389–420.

Wilson, D., Carroll, G.C. 1994. Infection studies of *Discula quercina*, an endophyte of *Quercus garryana*. *Mycologia* 86:635–647.

Wolverton, B.C. 2008. *How to Grow Fresh Air. 50 Houseplants that Purify Your Home and Office*. Weidenfeld & Nicolson, London, UK, 1997.

Worapong, J., Strobel, G., Ford, E.J. et al. 2001. *Muscodor albus* anam. gen. et sp. nov., and endophyte from *Cinnamomum zeylanicum*. *Mycotaxon* 79:67–79.

Worapong, J., Strobel, G.A., Daisy, B., Castillo, U., Baird, G., Hess, W.H. 2002. *Muscodor roseus* anna. nov. an endophyte from *Grevillea pteridifolia*. *Mycotaxon*, 81:463–475.

Wrigley, S.K., Sadeghi, R., Bahl, S., Whiting, A.J., Ainsworth, A.M., Martin, S., Katzer, W., Ford, R., Kau, D.A., Robinson, N., Hayes, M.A., Elcock, C., Mander, T., Moore, M.J. 1999. A novel (6S)-4,6- dimethyldodeca-2E,4E- dienoyl ester of phomalactone and related a- pyrone esters from a *Phomopsis* sp. with cytokine production inhibitory activity. *Journal of Antibiotics*, 52:862–872.

Wu, W., Davis, R.W., Tran-Gyamfi, M.B., Kuo, A., Labutti, K., Mihaltcheva, S., Hundley, H., Chovatia, M., Lindquist, E., Barry, K. 2017. Characterization of four endophytic fungi as potential consolidated bioprocessing hosts for conversion of lignocellulose into advanced biofuels. *Applied Microbiology and Biotechnology* 101(6):2603–2618. doi:10.1007/s00253-017-8091-1

Wu, W., Tran, W., Taatjes, C.A., Alonsogutierrez, J., Lee, T.S., Gladden, J.M. 2016. Rapid discovery and functional characterization of terpene synthases from four endophytic Xylariaceae. *PLoS ONE* 11(2):e0146983. doi:10.1371/journal.pone.0146983

Xia, X., Lie, T.K., Qian, X., Zheng, Z., Huang, Y., Shen, Y. 2011. Species diversity, distribution, and genetic structure of endophytic and epiphytic *Trichoderma* associated with banana roots. *Microbial Ecology* 61:619–625.

Xie, X.G., Fu, W.Q., Zhang, F.M., Shi, X.M., Zeng, Y.T., Li, H., Zhang, W., Dai, C.C. 2017. The endophytic fungus *Phomopsis liquidambari* increases nodulation and N2 fixation in *Arachis hypogaea* by enhancing hydrogen peroxide and nitric oxide signaling. *Microbial Ecology* 74(2):427–440.

Xiong, Z.Q., Yang, Y.Y., Na, Z., Yong, W. 2013. Diversity of endophytic fungi and screening of fungal paclitaxel producer from Anglojap yew, *Taxus x media*. *BMC Microbiology* 13(1):71. doi:10.1186/ 1471-2180-13-71

Xu, S., Ge, H.M., Song, Y.C., Shen, Y., Ding, H., Tan, R.X. 2009. Cytotoxic cytochalasin metabolites of endophytic *Endothia gyrosa*. *Chemical Biodiversity* 6:739–745. doi:10.1002/cbdv. 200800034

Yadav, R., Singh, A.V., Joshi, S., Kumar, M. 2015. Antifungal and enzyme activity of endophytic fungi isolated from *Ocimum sanctum* and *Aloe vera*. *African Journal of Microbiology Research* 9:1783–1788.

Yahara, I., Harada, F., Sekita, S., Yoshihira, K., Natori, S. 1982. Correlation between effects of 24 different cytochalasins on cellular structures and cellular events and those on actin in vitro. *Journal of Cell Biology* 92:69–78.

Yan, H.J., Gao, S.S., Li, C.S., Li, X.M., Wang, B.G. 2010. Chemical constituents of a marine-derived endophytic fungus *Penicillium commune* G2M. *Molecules* 15:3270–3275.

Yan, J.K., Ding, L.Q. Shi, X.L., Donkor, P.O., Chen, L.X., Qiu, F. 2017. Megastigmane glycosides from leaves of *Eucommia ulmoides* Oliver with ACE inhibitory activity. *Fitoterapia*, 116:121–125. doi:10.1016/j.fitote.2016.12.001

Yang, J.X., Chen, Y.G., Huang, C.H., She, Z.G. 2011. A new isochroman derivative from the marine fungus Phomopsis sp. (No. ZH-111). *Chemistry of Natural Compounds* 47:13. doi:10.1007/s10600-011-9820-9

Yang, Z.J., Ding, J.W., Ding, K.S., Chen, D.J., Cen, S., Ge, M. 2013. Phomonaphthalenone A: A novel dihydronaphthalenone with anti-HIV activity from *Phomopsis* sp. HCCB04730. *Phytochemistry Letters* 6(2):257–260. doi:10.1016/j.phytol.2013.02.003

Yedidia, I., Benhamou, N., Chet, I. 1999. Induction of defense responses in cucumber plants (*Cucumis sativus* L.) by the biocontrol agent *Trichoderma harzianum*. *Applied Environmental Microbiology* 65:1061–1070.

You, X., Feng, S., Luo, S.L. et al. 2013. Studies on a rhein-producing endophytic fungus isolated from *Rheum palmatum* L. *Fitoterapia* 85:161–168. doi:10.1016/ j.fitote.2012.12.010

Yu, B.Z., Zhang, G.H., Du, Z.Z., Zheng, Y.T., Xu, J.C., Luo, X.D. 2008. Phomoeuphorbins A-D, azaphilones from the fungus *Phomopsis euphorbiae*. *Phytochemistry* 69(13):2523–2526. doi:10.1016/j.phytochem.2008.07.013

Zahoor, M., Irshad, M., Rahman, H., Qasim, M., et al. 2017. Alleviation of heavy metal toxicity and phytostimulation of *Brassica campestris* L., by endophytic *Mucor* sp. MHR-7. *Ecotoxicology and Environmental Safety*, 142:139–149. https://doi.org/10.1016/j.ecoenv.2017.04.005

Zeng, Y., Wang, H., Kamdem, R.S.T., Orfali, R.S., Dai, H., Makhloufi, G., Janiak, C., Liu, Z., Proksch, P. 2016. A new cyclohexapeptide, penitropeptide and a new polyketide, penitropone from the endophytic fungus *Penicillium tropicum*. *Tetrahedron Letters* 57:2998–3001.

Zhang, D.W., Tao, X.Y., Liu, J.M., Chen, R.D., Zhang, M., Fang, X.M., Yu, L.Y., Dai, J.G. 2016b. A new polyketide synthase–nonribosomal peptide synthetase hybrid metabolite from plant endophytic fungus *Periconia sp. Chinese Chemistry Letters* 27:640–642.

Zhang, H., Ruan, C., Bai, X. 2015. Isolation and antimicrobial effects of endophytic fungi from *Edgeworthia chrysantha. Bangladesh Journal of Pharmacology* 10:529–532.

Zhang, H., Sun, X., Xu, C. 2016a. Antimicrobial activity of endophytic fungus *Fusarium* sp. Isolated from medicinal honeysuckle plant. *Archives Biological Science* 68:25–30.

Zhang, X.Y., Liu, X.Y., Liu, S.S. et al. 2011. Responses of *Scirpus triqueter*, soil enzymes and microbial community during phytoremediation of pyrene contaminated soil in simulated wetland. *Journal of Hazardous Materials* 193:45–51. doi:10.1016/j.jhazmat.2011.07.094

Zhang, Y., Liu, S., Che, Y. et al. 2007. Epicoccins A-D, epipoly-thiodioxopiperazines from a Cordyceps-colonizing isolate of *Epicoccum nigrum. Journal of Natural Product* 70:1522–1525.

Zhao, J., Wang, Z., Dai, Y., Xing, B. 2013. Mitigation of CuO nanoparticle-induced bacterial membrane damage by dissolved organic matter. *Water Research*, 47(12):4169–4178. https://doi.org/10.1016/j.watres.2012.11.058

Zhao, K., Ping, W., Zhang, L., Liu, J., Lin, Y., Jin, T., et al. 2008. Screening and breeding of high taxol producing fungi by genome shuffling. *Science in China Series C: Life Sciences*. 51:222–231. https://doi.org/10.1007/s11427-008-0037-5

Zhong, L.Y., Zou, L., Tang, X.H., Li, W.F., Li, X., Zhao, G., Zhao, J.L. 2017. Community of endophytic fungi from the medicinal and edible plant *Fagopyrum tataricum* and their antimicrobial activity. *Tropical Journal of Pharmaceutical Research* 16(2):387–396. doi:10.4314/tjpr.v16i2.18

Zikmundova, M., Drandarov, K., Bigler, L., Hesse, M., Werner, C. 2002. Biotransformation of 2-Benzoxazolinone and 2-Hydroxy-1,4-Benzoxazin-3-one by endophytic fungi isolated from *Aphelandra tetragona. Applied Environmental Microbiology* 68(10):4863–4870. doi:10.1128/AEM.68.10.4863-4870.2002

Zou, W.X., Meng, J.C., Lu, H. et al. 2000. Metabolites of *Colletotrichum gloeosporioides*, an endophytic fungus in *Artemisia mongolica. Journal of Natural Product* 63:1529–1530.

Zuccaro, A., Lahrmann, U., Guldener, U., Langen, G., Pfiffi, S., Biedenkopf, D., Wong, P., Samans, B., Grimm, C., Basiewicz, M., Murat, C., Martin, F., Kogel, K.-H. 2011. Endophytic life strategies decoded by genome and transcriptome analyses of the mutualistic root symbiont *Piriformospora indica. PLoS Pathogens* 7: e1002290. doi:10.1371/journal.ppat.1002290

Index

Note: Page numbers in bold and italics refer to tables and figures, respectively.

A

abiotic factors, 2
Acanthus ilicifolius, 20
agriculture applications, 22–23, **24–27**
anthropogenic perturbations, 14
anticancer compounds, 30, 36
antimicrobial compounds, 28–30, **31–35**
antitumor applications, 36, **37–38**

B

bacterial endophytes, 1
bioenergy and biocatalysis applications, 48–49
BioEnsure R-*Corn*, 64
biofertilizers, 16
bioformulations, 63, *64*
biomedicine-related applications, 39, **40**
biotransformation reactions, 39, 41–42

C

chemical/heat treatments, 53
chitinase enzyme, 45
citrinin, 45
co-culture fermentation, 67–68
commercialization, 63–64
compounds/metabolites, 36, 39
Curvularia lunata, 39, 41
cytoskyrins, 58

D

Diaporthe and *Phomopsis* species, 58–59
Dicerandrol D, 58
dicerandrols, 58–59
diversity and ubiquity, 1–4
drought resistance, 17

E

endophytes, xi
  challenges and limitations, 65–69
  communities in tropics, 14–15
  and endophyte-host plant association, 15–18
  overview, xvii–xx
  research, 18–20
  and tropical host plants, 4, **5–10**, *11–12*, 12–13
endophytic actinobacteria, 57–58
endophytic fungi, 23
enzymes, 43–45
Epicoccum nigrum, 46
exopolysaccharides (EPS), 36

F

fertilizer application, 15
fibrinolytic enzyme, 45
functional polysaccharides, 36
fungal endophytes, 1, 18, 29
fungal species, 19, *19*

G

generally regarded as safe (GRAS) compounds, 30
genetic engineering, 69

H

HAQN (hydroxyanthraquinoid) pigments, 45
hennotannic acid, 46
horizontal-transfer, 15
host generalism, 4
hydrocarbon bioremediation, 54–55
hydroxyanthraquinoid (HAQN) pigments, 45
hyperaccumulator plants, 49–50

I

indole-3-acetic acid (IAA), 51
industrial applications
  enzymes, 43–45
  pigments/colorants, 45–46, **47–48**

L

laccase enzyme, 44
Lasiodiplodia species, 59–60
L-asparaginase, 30
lawsone (2-hydroxy-1, 4-naphthoquinone), 46
liquid formulation, 63
lovastatin compound, 39

M

Makassar fruit plant, 18
metal bioremediation process, 49–54
metal pollutants, 49
Monascus species, 45
Muscodor species, 60
mycodiesel producer, 48
mycoparasite, 60
mycorrhizal endophytes, 1

N

nanoparticles (NPs), 42–43
nanotechnology, 42
natural perturbations, 14
NPs (nanoparticles), 42–43

O

Ocimum sanctum, 20

P

passive process, 52
pathogens, 13
Phomopsis species, 29
phomoxanthones compounds, 59
phosphate solubilization, 51
phosphate solubilizers, 16

pigments/colorants, 45–46, **47–48**
pinoresinol diglucoside compound, 39
Piriformospora indica, 51, 63
polyester polyurethane (PUR), 55
polyketide pigments, 45
protease, 44
Pseudomonas species, 29

Q

quinones, 59

R

Rafflesiaceae members, 13

S

sampling procedure, 65–66
siderophore production, 51
silver nanoparticles (AgNPs), 42
soil-inhabiting microbes, 42
Solanum trilobatum, 20

T

Talaromyces species, 46
tannase, 43
taxol/paclitaxel, 21
terpenoids, 59
Trichoderma species, 60–61
Tropic of Cancer, xviii, *xix*
Tropic of Capricorn, xviii, *xix*

V

vertical-transfer, 15
vincristine compound, 41
Viscum coloratum, 59
volatile compounds, 30, 39
volatile organic compounds (VOCs), 48

X

xenobiotic pollutants, 55
Xylaria species, 61–62

Printed and bound by CPI Group (UK) Ltd, Croydon, CR0 4YY

24/10/2024

01778280-0001